# MINI
# WEAPONS
## OF MASS DESTRUCTION®
## BUILD AND MASTER NINJA WEAPONS

## JOHN AUSTIN

CHICAGO
REVIEW
PRESS

**Library of Congress Cataloging-in-Publication Data**

Austin, John, 1978–

   Miniweapons of mass destruction. Build and master ninja weapons / John Austin.

      pages cm

   Summary: "Author and toy designer John Austin provides detailed, step-by-step instructions with diagrams to show stealth warriors how to build 37 different ninja weapons for the modern era. Each of the projects in *MiniWeapons of Mass Destruction: Build and Master Ninja Weapons* is built from common household and office items—plastic utensils, markers, clothespins, paper clips, wire hangers, and discarded packaging—all clearly detailed on materials lists. Builders are offered a variety of samurai stars, blowguns, throwing darts, siege weapons, and ninja tools to choose from. Once they've assembled their armory, the author provides novices several targets to construct to practice their shooting skills. Armed, trained, and shrouded in black, readers are now prepared for missions of reconnaissance, sabotage, and other grim errands" – Provided by publisher.

   ISBN 978-1-61374-924-1 (paperback)

   1. Martial arts weapons. 2. Handicraft. 3. War toys. 4. Ninja. 5. Ninjutsu. I. Title. II. Title: Build and master ninja weapons. III. Title: Mini weapons of mass destruction, build and master ninja weapons.

   GV1101.5.A97 2014

   796.8–dc23

                              2014021512

Cover and interior design: Jonathan Hahn

Illustrations: Austin Design, Inc.

© 2014 by Austin Design, Inc.

All rights reserved

Published by Chicago Review Press, Incorporated

814 North Franklin Street

Chicago, Illinois 60610

ISBN 978-1-61374-924-1

Printed in the United States of America

10 9 8 7 6 5 4 3

**Dedicated to my *sensei no haha* (sensei mother), for cutting a path of creativity for me!**

Ninja wisdom holds that revenge is the dream of the weak, and a disciplined ninja knows that no enemy is forever. If you feel the desire for vengeance stirring within you, do not succumb. It is said that the perfect master avoids the fight and the very best swords are kept in their sheaths.

**Join the MiniWeapons army on Facebook:**
MiniWeapons of Mass Destruction: Homemade Weapons Page

**For video demos, tutorials, and other extras, find us on YouTube:**
MiniWeaponsBook Channel

# CONTENTS

# INTRODUCTION

Fill your ninja dojo with *MiniWeapons of Mass Destruction 4*, a home-made weapons guide that will help you transform everyday items into specialized ninja weapons.

Cloaked head to toe in black to conceal his or her identity, a ninja needs a small arsenal of weapons and secret devices for sabotage, espionage, reconnaissance, infiltration, and other grim missions. You trained in the art of *ninjutsu* but will also need to master ranged weapons to avoid detection. Lucky for you, those weapons are heavily represented in this book.

One needs time to master the art of stealth and invisibility, so building these ranged and melee weapons should be fast! To speed the process, each project in this book is accompanied by an easy-to-read bill of materials, step-by-step instructions, and alternative construction methods. The final chapter offers a small library of simple ninja targets, perfect for testing the homemade weaponry.

This book is for all ninja practitioners, brand-new and longtime masters. It pushes the laws of physics with basic engineering, inspires creativity, proposes experimentation, and fuels the imagination with material exploration. Most of the ninja weapons are great representations of their real-life counterparts, but they cost little to no yen (money), making them great for group exercises and perfect for arming a ninja clan.

Keep in mind that this book is for entertainment purposes only. **Please review the "Play it Safe" section for your personal protection.** Build and use these projects at your own risk.

# PLAY IT SAFE

Crafting and mastering ninja weapons requires discipline; even an innocent-looking ninja is always contemplating mayhem. When building and firing *MiniWeapons*, be responsible and take every safety precaution. Switching materials, substituting ammunition, assembling improperly, mishandling, targeting inaccurately, and misfiring can all cause harm. Like any trained ninja, you should always be prepared for the unknown. **Eye protection is a must** if you choose to experiment with any of these projects.

Always be aware of your environment, including spectators. The suggested blowgun designs are capable of firing with incredible range and accuracy, and can cause harm if misused. Throwing darts and throwing stars have sharp points and should be used in a responsible manner. Some of the elastic siege weapons launch projectiles with unbelievable force and can cause damage, so be careful when operating them. Ammo, no matter what the material, can cause eye damage. **Never point these launchers at people, animals, or anything of value.** And **never** take or transport any of these projects on public transportation, such as an airplane, bus, or train—these projects are to be used at home.

Remember that because miniweaponry is homebuilt, it is not always accurate. Basic target blueprints and proposed printouts are available at the end of the book and at www.JohnAustinBooks.com. Use these—not random targets—to test the accuracy of your MiniWeapon.

Some of the projects outlined in this book require tools such as hobby knives, pocketknives, hot glue guns, wire cutters, and electric drills, which can cause injury if handled carelessly. Tools need your full attention—make safety your number-one priority. If you have trouble cutting, your knife may be dull or the selected material may be too hard; stop immediately and substitute one of the two. ***Junior ninja practitioners should always be assisted by an adult sensei when handling potentially harmful tools.***

Always be responsible when constructing and using miniweaponry. It is important that you understand that the author, the publisher, and the bookseller cannot and will not guarantee your safety. When you try the projects described here, you do so at your own risk. They are *not* toys!

I, _____, understand that safety is my number-one priority.
　　　(NINJA'S NAME)

# THROWING STARS

# CARDBOARD THROWING STAR

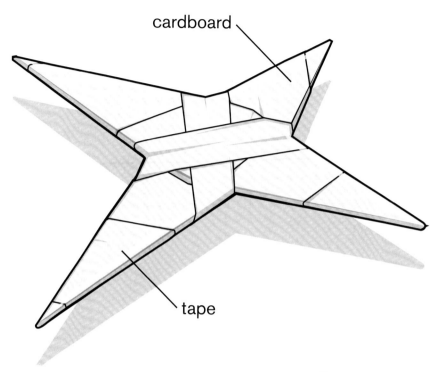

cardboard

tape

**Range: 10–30 feet**

The Cardboard Throwing Star is perhaps the most popular MiniWeapon in the young ninja's arsenal due to its straightforward construction and easily accessible materials. The design is innocent enough for indoor dojo training, and if you are a sensei on a budget, this ninja star costs very few yen to mass produce.

## Supplies

1 piece of corrugated card-
    board
Metallic duct tape (or similar)

## Tools

Safety glasses
Scissors
Pen
Ruler

# Step 1

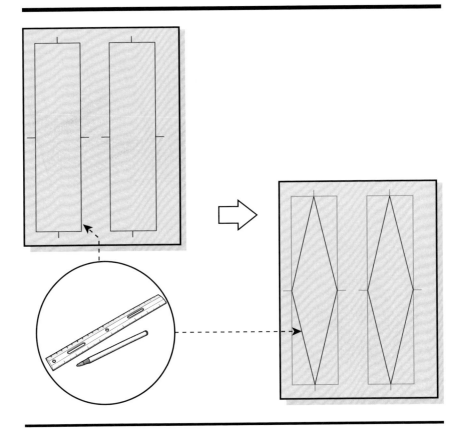

Begin construction by cutting a piece of corrugated cardboard to be approximately 8 inches by 4 inches. Corrugated cardboard from a shipping box is ideal because of its added weight, which will increase the star's throwing range.

With a pen and ruler, draw two identical rectangles, 6 inches long by 1½ inches wide, on the cardboard piece. Then on both rectangles, make a small mark at the halfway point on each of the four sides. Draw a straight line to each halfway mark, creating two diamonds as shown.

# Step 2

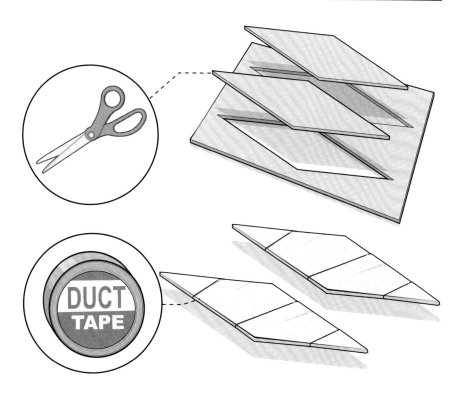

Use scissors to cut both diamonds from the cardboard. Discard the leftover cardboard scrap.

Wrap both cardboard diamonds with metallic duct tape. The tape's finish will give this throwing star an authentic metal appearance. However, electrical and masking tape will also work.

# Step 3

Stack both diamonds together at their center points, as shown. Secure the sections by wrapping tape around the four corners of the assembly.

The Cardboard Throwing Star can be thrown horizontally or vertically using a motion similar to that for throwing a Frisbee. Hold the star parallel to your palm, gripping it between your thumb and index finger. Then use the weight of your body and arm in tandem to add power, letting the star slip from your fingertips. Be prepared: hold a stack of throwing stars in your opposite hand that can be slid off and flicked in rapid succession. Test your ninja star throwing skills with the Lantern Knockout (page 289).

**Remember to use eye protection!** Cardboard Throwing Stars have pointed tips and can reach high speeds. **MiniWeapon projects are not meant for use on living targets.** Always stay clear of spectators and throw in a controlled manner.

# PAPER PLATE NINJA STAR

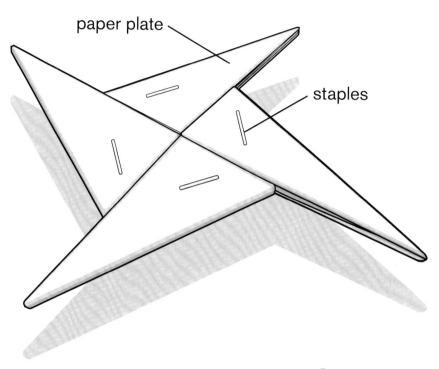

paper plate

staples

**Range: 10–30 feet**

Gripping the Paper Plate Ninja Star, an innocent-looking ninja is contemplating mayhem. Constructed from a single paper plate and a few slices of the *katana* (sword), the star can quickly be folded and assembled into a formidable secondary weapon. Plus, with plates available in bundles of 50 or more, you can afford to outfit a whole clan of ninjas.

## Supplies

1 disposable paper plate (thin, round)

## Tools

Safety glasses
Marker
Ruler
Drinking glass, plastic cup, or plastic lid (approximately 2½-inch diameter)
Scissors
Stapler
Clear tape

# Step 1

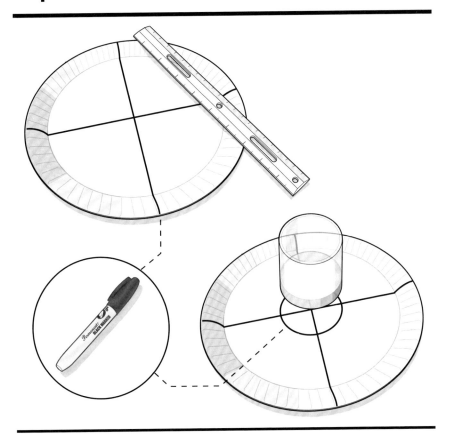

For this project to be successful, use a very thin, round disposable paper plate. Place the plate face up and locate the approximate center point. Use a marker and a ruler to divide the plate into four equal parts by drawing two guidelines from edge to edge as shown.

Next, place the glass, cup, or lid in the center of the plate and trace the circle diameter onto the plate with the marker. This added center circle will be the core of the throwing star.

# Step 2

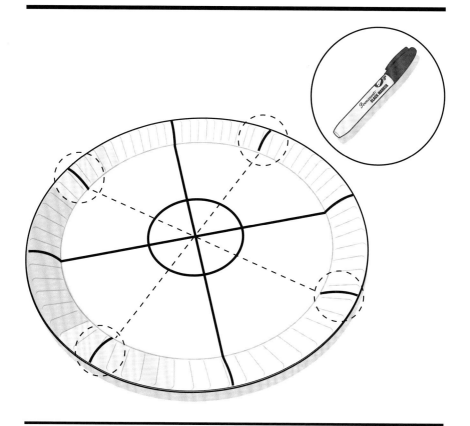

Divide the four sections into eight sections by placing the ruler as shown by the dashed guidelines. With the ruler intersecting the marked center point and positioned in the middle of the guidelines from step 1, mark two solid lines at opposite ends, starting from the plate's edge. Complete this step once more until the plate's outer edge is divided into eight equal parts.

# Step 3

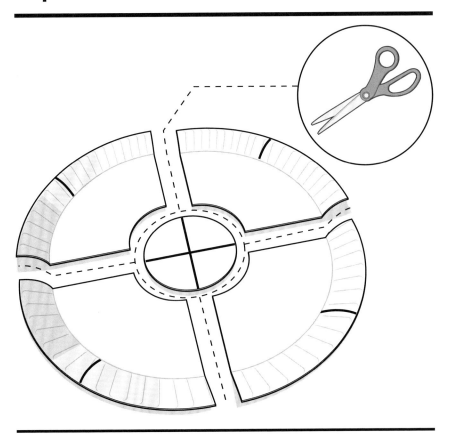

With scissors, cut along the large guidelines, stopping at the drawn center circle. Then complete the cut by going around the center circle as shown. **Do not** cut the 1-inch guidelines from step 2. When finished, the plate should be divided into five sections, with one of the sections being the center circle.

# Step 4

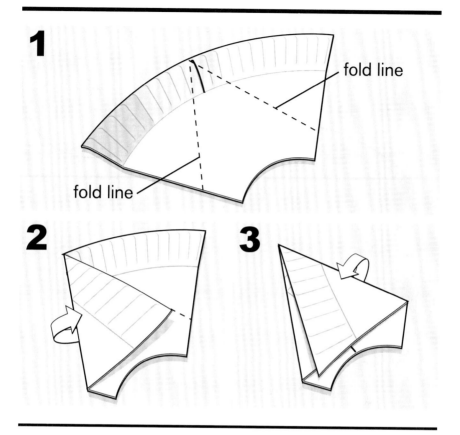

**1** fold line

fold line

**2**

**3**

Each wedge will be transformed into one of the four points of the ninja star. Start by locating the center guideline you marked in step 2. The next two fold lines will extend from this line, also known as the vertex, as indicated by the dashed lines shown (1).

Using the marked line as the starting point, fold the right side of the plate to create an overlapping triangle (2). Then repeat this fold on the left side, with an identical triangle, overlapping the first folded triangle (3). When you're finished, both folds should create a triangular shape with two sides equal in length (isosceles triangle).

# Step 5

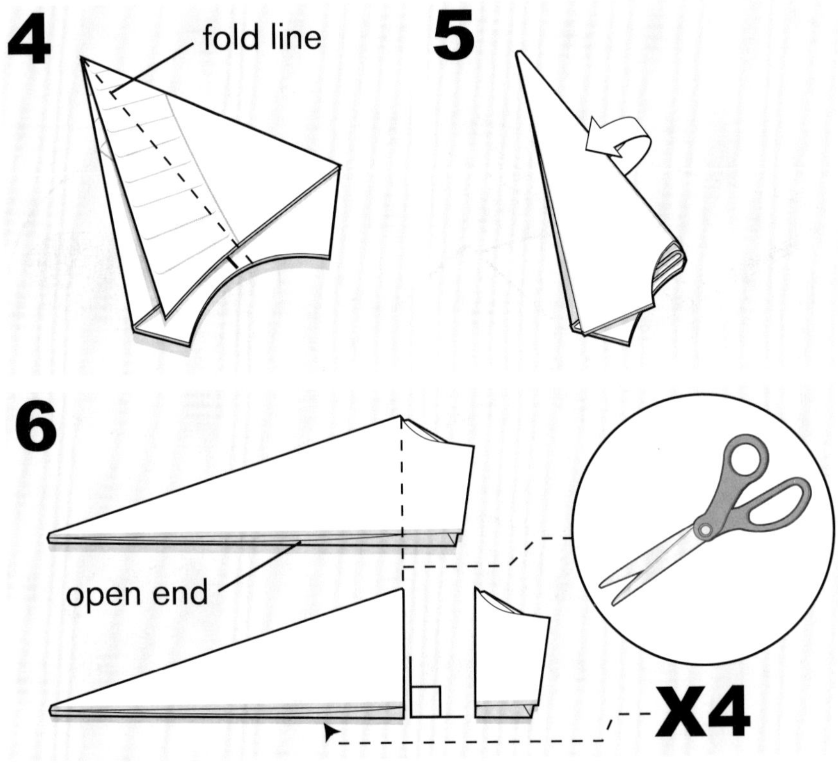

**4** fold line

**5**

**6** open end

**X4**

The ninja points will be completed with one additional fold down the center (the altitude). Using the dotted line shown as the crease line, fold the paper assembly in half (4 and 5).

Position the paper wedge horizontally, with the open side of the wedge down, as shown. With scissors, trim off the back detail to create a right triangle (6). Repeat the folding and trimming for the remaining three wedges.

# Step 6

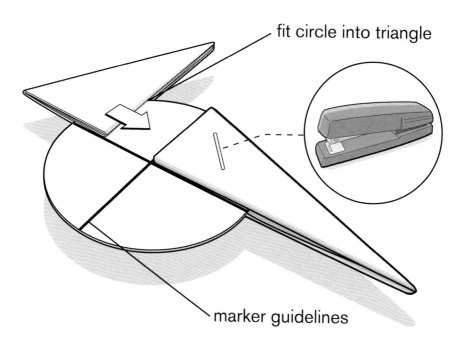

fit circle into triangle

marker guidelines

The small circle you created in step 3 is the core of the Paper Plate Ninja Star. To add the star points, slide one of the paper wedges over the paper circle, then align the right triangle with the drawn marker guidelines on the circle. Once in place, secure the ninja point with a stapler.

Add another ninja point to the assembly by aligning the 90-degree corner of the second wedge against the mounted ninja point, while keeping within the marker guidelines. Staple the wedge in place.

# Step 7

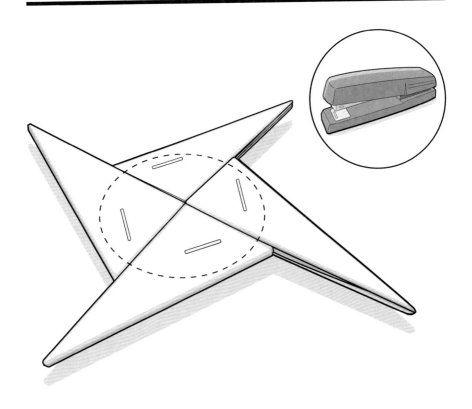

Slide the remaining two ninja points onto the circle base and secure them with additional stapling. The Paper Plate Ninja Star is complete! If you are concerned about the sharp points of the added staples, cover the metal edges with clear tape for added safety.

A throwing star can be thrown horizontally or vertically using a motion similar to that for throwing a Frisbee. Hold the star parallel to your palm, gripping it between your thumb and index finger. Then use the weight of your body and arm in tandem to add power, letting the star slip from your fingertips. Be prepared: hold a stack of throwing stars in your opposite hand that can be slid off and flicked in rapid succession.

*Remember to use eye protection!* Throwing stars have pointed tips and can reach high speeds. *MiniWeapon projects are not meant for use on living targets.* Always stay clear of spectators and throw in a controlled manner.

# PAPER TUBE NINJA STAR

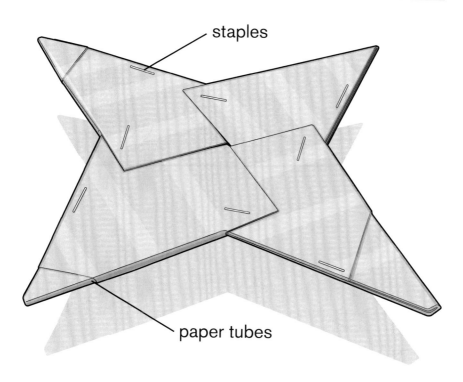

staples

paper tubes

**Range: 10–30 feet**

A short distance from the main tower, you move stealthily to within sight of the enemy, cloaked in the shadows of the fortification. Your recon mission is complete, and with a quick diversion, you can slip away unnoticed. The Paper Tube Ninja Star is the perfect choice for this task because of the star's commanding size. When thrown in the vicinity of any imperial soldier, your opponent will be distracted just long enough for you to exit.

## Supplies

4 toilet paper tubes (or similar)

## Tools

Safety glasses
Scissors
Stapler
Clear tape

# Step 1

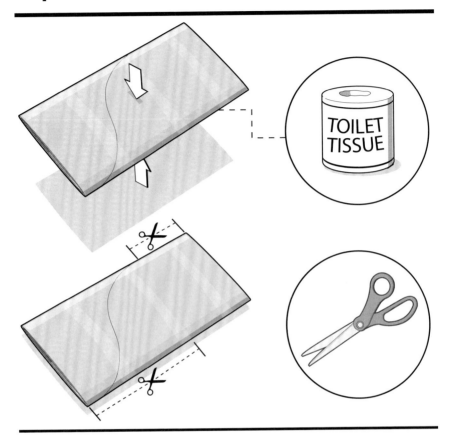

This ninja project requires four toilet paper tubes (or similar tubes approximately 4 inches long). Flatten all four tubes, and crease each edge to transform them into doubled-walled rectangles, as illustrated.

With one tip of a pair of scissors inside the cardboard tube, cut a 3-inch slit along one crease. On the opposite side of the tube, cut a 1-inch slit along the crease. Repeat both cuts on the three remaining tubes.

# Step 2

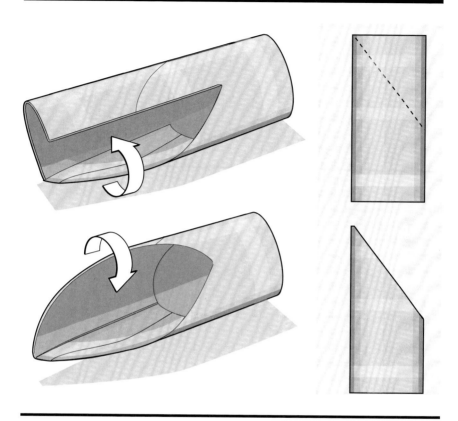

Using the cut 3-inch slit, fold both corresponding sides into the tube to create a point at the end of the tube as shown.

The folded-in edges should align, as illustrated in the bottom left image. When the fold is complete, press the tube flat so the finished wedge looks like the bottom right image, and repeat this step for the remaining three tubes.

# Step 3

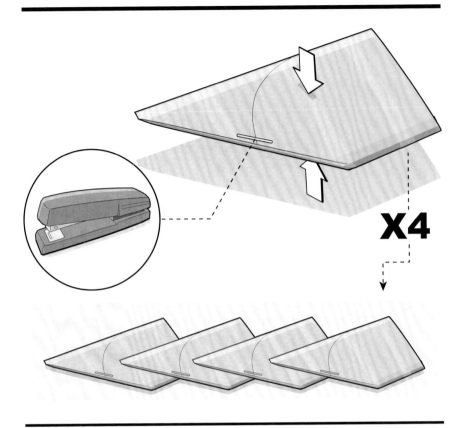

With the tube flattened and the edges creased, add one staple to the angled side, centered in the middle of the slant. Repeat this step for the three remaining tubes. Each modified tube will become one ninja point on the throwing star.

# Step 4

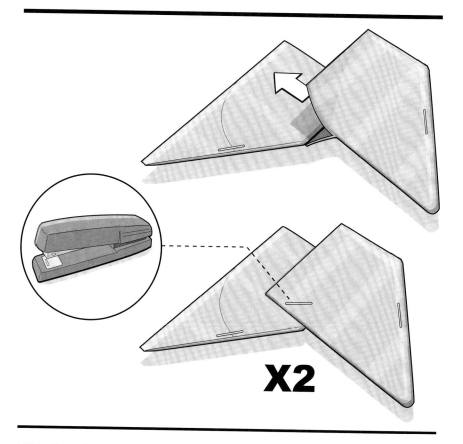

**X2**

With the tube angles facing counterclockwise, connect two tubes by sliding the first tube 90 degrees into the second tube's 1-inch pocket. When the first tube's angle touches the second tube's 1-inch opening, staple both together.

Repeat this assembly with the remaining two tubes. When you're finished you will have two assemblies that are exactly the same.

# Step 5

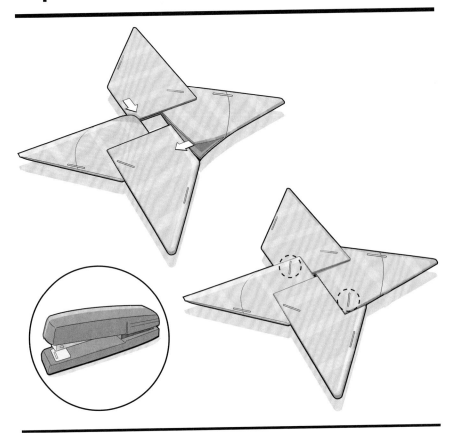

Fit both assemblies together by sliding each uncut end into the 1-inch slits on the opposite assembly. Slide the sections together until they are snug, then use a stapler to secure the blades. The Paper Tube Ninja Star is now complete and ready for a test flight. You can place clear tape over the metal staple points for added safety.

This large throwing star can be thrown horizontally or vertically using a motion similar to that for throwing a Frisbee. Hold the star parallel to your palm, gripping it between your thumb and index finger. Then use the weight of your body and arm in tandem to add power, letting the star slip from your fingertips. A stack of additional stars in your opposite hand can be slid off and flicked in rapid succession.

**Remember to use eye protection!** Throwing stars have pointed tips and can reach high speeds. **MiniWeapon projects are not meant for use on living targets.** Always stay clear of spectators and throw in a controlled manner.

# TRI-POINT NINJA STAR

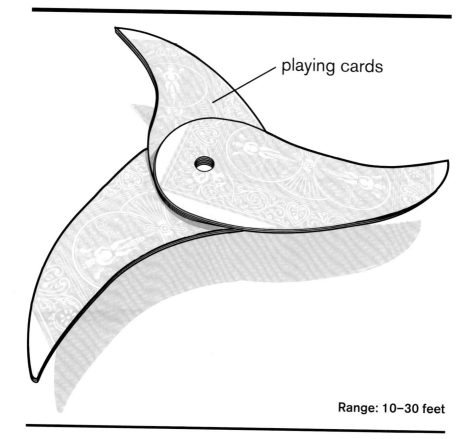

playing cards

**Range: 10–30 feet**

Espionage, sabotage, and infiltration are just some of the dangerous games ninjas play. So when it's your turn to deal up some misfortune, stock up on a few Tri-Point Ninja Stars. Constructed from a short stack of laminated playing cards, this weapon's simplicity will impress your ninja clan. Its durable design has an impressive range that will intimidate any betting foot soldier. Plus, the blade design and the point numbers can be customized for mission-specific uses.

## Supplies

9 playing cards

## Tools

Safety glasses
Glue stick

Marker
Plastic milk carton cap
 (optional)
Scissors
Single-hole punch
Hot glue gun

# Step 1

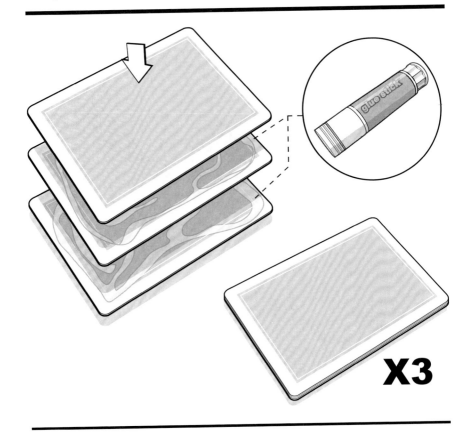

Stack three playing cards, aligning their edges. Use a glue stick to attach them. (Hot glue, white glue, or superglue will also work if a glue stick is not available.) Repeat this step with the remaining playing cards to make two additional sets of three. You'll have a grand total of three combined sets, and each set will make up one ninja point. Let all three sets dry completely before proceeding to the next step.

# Step 2

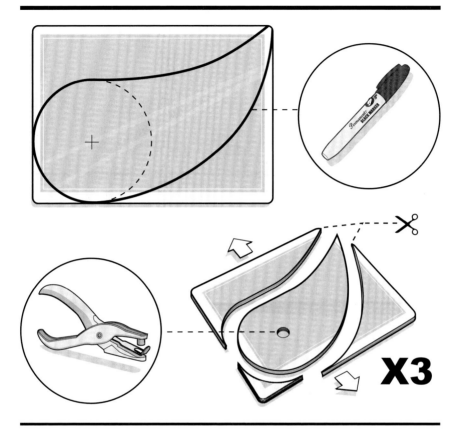

Starting with one set of cards, use a marker to draw a circle in the lower left corner, similar to the pattern shown. For a perfect circle (optional), trace the diameter of a plastic milk carton cap. Finish the ninja point pattern by extending two lines from the top and bottom of the circle to the opposite corner in a curved blade design, as shown. The blade design can be unique, so have fun with it!

Cut out the pattern with scissors. Then, with a single-hole punch, make one hole in the center of the initial circle. This hole will help you align all three finished blades, and it will allow you to string or clip multiple finished blades together when transporting them.

Trace the first blade pattern onto the remaining two sets of cards, so when all three blade patterns are cut out and hole punched, they are identical.

# Step 3

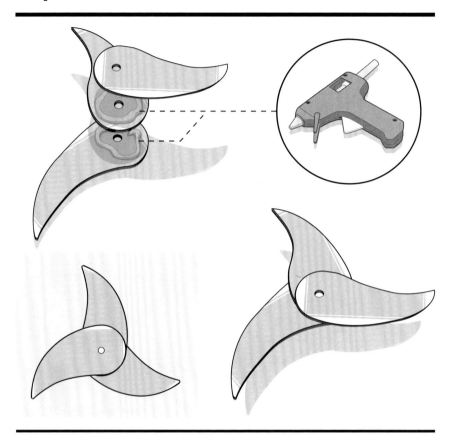

Stack and evenly space the blades as shown—with the hole in the center—and hot glue the assembly together. Allow the throwing star to dry prior to testing.

The Tri-Point Ninja Star can be thrown horizontally or vertically using a motion similar to that for throwing a Frisbee. Hold the star parallel to your palm, gripping it between your thumb and index finger. Then use the weight of your body and arm in tandem to add power, letting the star slip from your fingertips. Be prepared: hold a stack of throwing stars in your opposite hand that can be slid off and flicked in rapid succession. Test your ninja star throwing skills with the Lantern Knockout (page 289).

**Remember to use eye protection!** Throwing stars have pointed tips and can reach high speeds. **MiniWeapon projects are not meant for use on living targets.** Always stay clear of spectators and throw in a controlled manner.

# GOLF TEE SHURIKEN

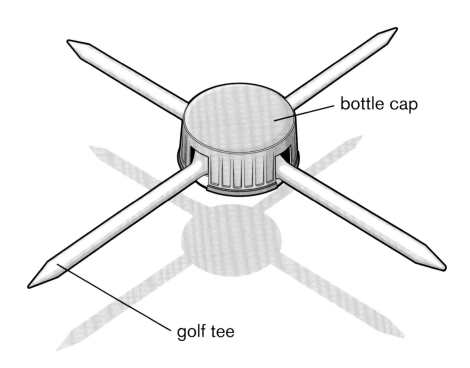

bottle cap

golf tee

**Range: 10–30 feet**

The Japanese word for throwing star is *shuriken*, which means "sword hidden in the hand." Historically, shuriken were crafted from a variety of everyday objects, just like this unforgiving Golf Tee Shuriken is. Assembled from wooden golf tees and a couple of plastic bottle caps, it sports multifaceted points that can be devastatingly lethal to cardboard! Although designed as a secondary weapon, this throwing star may become the most popular element of your ninja arsenal.

## Supplies

1 large plastic bottle cap
1 small plastic bottle cap
4 golf tees

## Tools

Safety glasses
Marker
Wire cutters
Hot glue gun

# Step 1

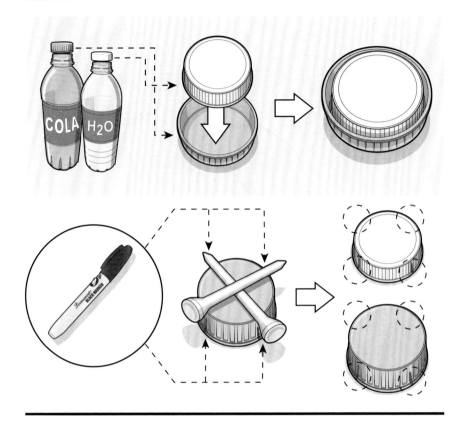

Locate two different-sized plastic bottle caps that fit smoothly within one another, as illustrated. The caps from a 16-ounce soft drink bottle and a 16-ounce water bottle work well.

Use a marker to indicate four even quadrants on both caps. As shown, you can use the golf tees as a straightedge if needed.

# Step 2

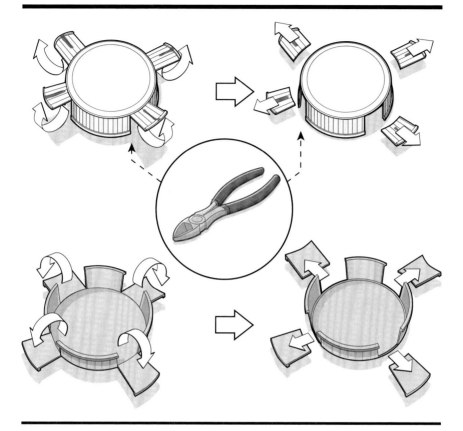

Use wire cutters to snip out a section of cap equivalent to the width of the golf tee, using each marked line from step 1 as a center point. Do this four times per cap, on both caps.

Once all eight sections are snipped, bend down the cap flaps and then use the wire cutters to remove the flap material from the caps. The finished caps should have four identical openings.

# Step 3

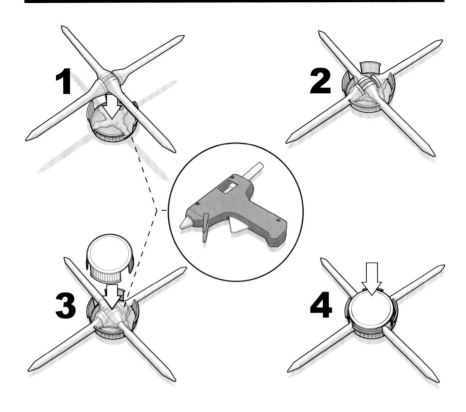

Carefully add hot glue to the inside of the largest cap (1), then arrange all four golf tees inside the cap with each tee pointing outward (2). Once in place, add additional hot glue over the four fixed golf tees (3), then place the smaller cap over the assembly, with the cap notches fitting around the attached golf tees. Depending on the amount of hot glue originally added, you might need to use the wire cutters to remove more material from the smaller cap for a snug fit.

The Golf Tee Shuriken can be thrown horizontally or vertically using a motion similar to that for throwing a Frisbee. **Remember to use eye protection!** The Golf Tee Shuriken has piercing tips and can reach high speeds. **MiniWeapon projects are not meant for use on living targets.** Always stay clear of spectators and throw in a controlled manner.

# 14-POINT THROWING STAR

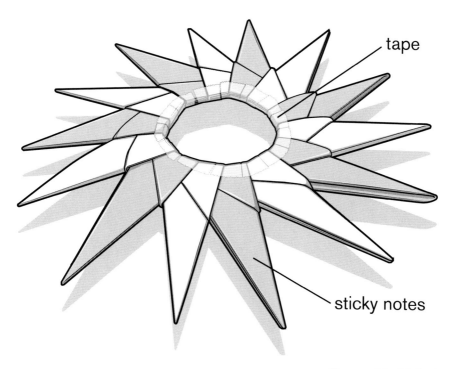

tape

sticky notes

**Range: 10–30 feet**

Once a closely guarded secret of cloak-and-shadow ninjas, here now is the 14-Point Throwing Star! Built around the idea of 14 rays of light radiating from the sun, this disciplined version is crafted from less lethal sticky notes. The center hole makes this MiniWeapon more aerodynamic—and easier for a ninja to carry on a string or belt.

## Supplies

14 square sticky notes (3 inches by 3 inches)
Clear tape

## Tools

Safety glasses

# Step 1

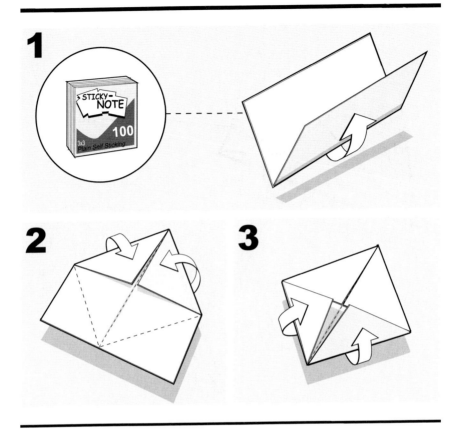

Fourteen square sticky notes will be used to construct the 14-Point Throwing Star. To make the star stand out, use two different colors of sticky notes. If sticky notes are not available, 14 squares of paper, 3 inches by 3 inches, can be substituted.

Start by folding one sticky note in half as shown, edge to edge (1). Then unfold that sticky note. Use the center crease line as a guide and fold over two corners so that the edges come together at the center crease but do not overlap (2). Repeat these folds for the remaining two corners; do not overlap any of the triangles (3). At the end, the modified paper should be a smaller square or triangle, depending on its orientation.

# Step 2

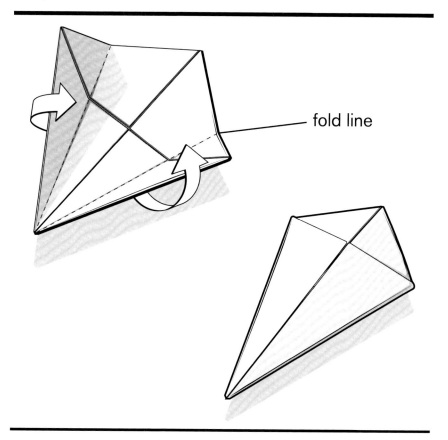

fold line

Fold inward two opposite sides of the triangle, resting the edges on the centerline, creating a diamond "kite" shape. Press your fingernail along the crease to hold the fold in place.

# Step 3

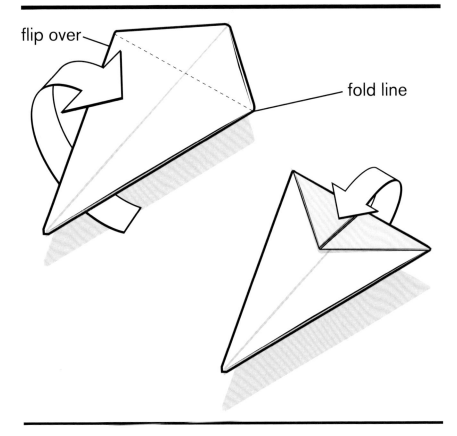

flip over

fold line

Now flip the paper diamond over so that the solid side is facing up. Fold the small end of the diamond "kite" onto the larger end to create a triangle.

# Step 4

With the small triangle folded onto the top surface, fold the paper assembly in half, down the center of the large triangle, keeping the small triangle on the outside as shown. This assembly is now finished, and will make one point of the throwing star.

# Step 5

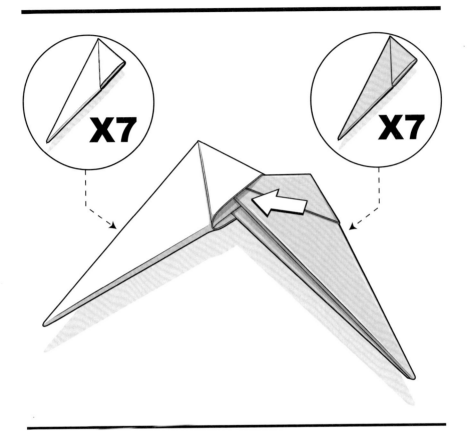

Repeat steps 1 through 4 with the remaining sticky notes until you have made 14 folded points, 7 of each color.

Start assembling the star with just two finished points. With both points positioned the same, slide the two paper assemblies together by sliding the lower point of the first assembly into the back pocket of the second assembly.

# Step 6

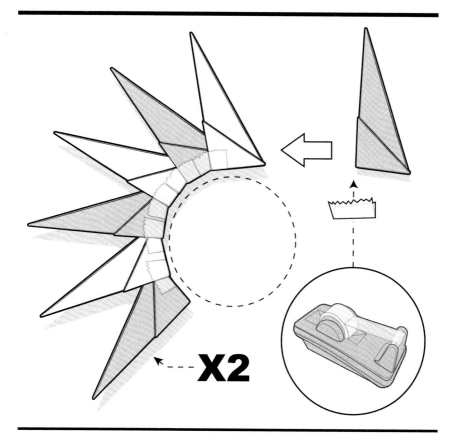

Continue to connect the paper assemblies seven more times, alternating colors, until the grouping represents a half arc as shown. Add clear tape to the inside of the arc to secure the assembly together.

Once you've finished with the above step, create another seven-point assembly, the same as the first.

# Step 7

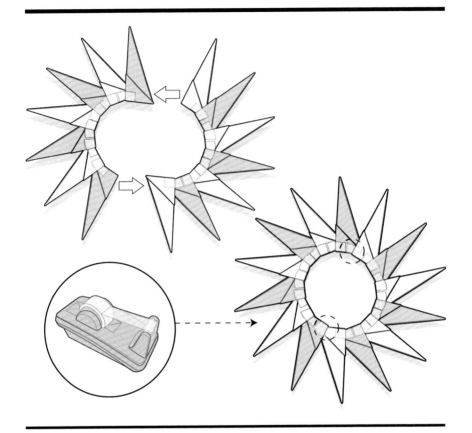

Combine both assemblies by tucking the ends into the corresponding pockets. Use tape to hold them together.

The 14-Point Throwing Star can be thrown horizontally or vertically using a motion similar to that for throwing a Frisbee. Hold the star parallel to your palm, gripping it between your thumb and index finger. Then use the weight of your body and arm in tandem to add power, letting the star slip from your fingertips. Be prepared: hold a stack of throwing stars in your opposite hand that can be slid off and flicked in rapid succession. Test your ninja star throwing skills with the Lantern Knockout (page 289).

***Remember to use eye protection!*** Throwing stars have pointed tips and can reach high speeds. ***MiniWeapon projects are not meant for use on living targets.*** Always stay clear of spectators and throw in a controlled manner.

# CD MANJI SHURIKEN

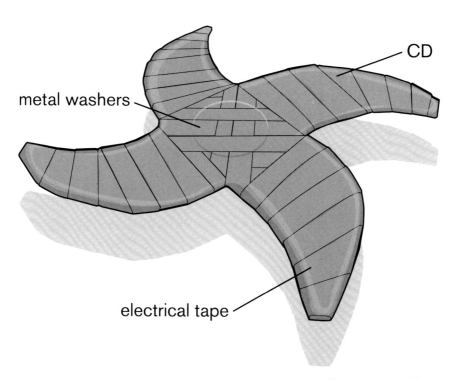

CD

metal washers

electrical tape

**Range: 10–30 feet**

The CD Manji Shuriken is a "hooked style" (*manji*) throwing star with four curved corners. Its design makes it versatile and easy to throw, which will be of great benefit to those still mastering *shurikenjutsu* (ninja star training). All throwing stars are utilized as secondary weapons to the more commonly used katana sword; however, the CD Manji Shuriken will certainly play a pivotal role in battle.

## Supplies

1 unwanted CD or DVD
2 metal washers or coins
Electrical tape or duct tape

## Tools

Safety glasses
Marker
Square sticky note (3 inches by 3 inches)
Large scissors
Hot glue gun

# Step 1

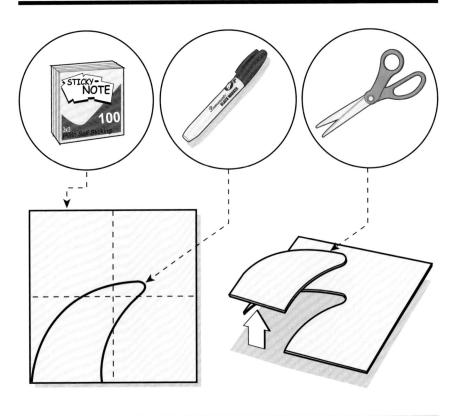

With a marker, draw a blade pattern on a 3-inch-by-3-inch sticky note, slightly larger than half the height and width of the note, using the illustration above as a guide. You can make center folds on the sticky note to help you stay within the general shape.

Use scissors to cut out the drawn blade pattern.

# Step 2

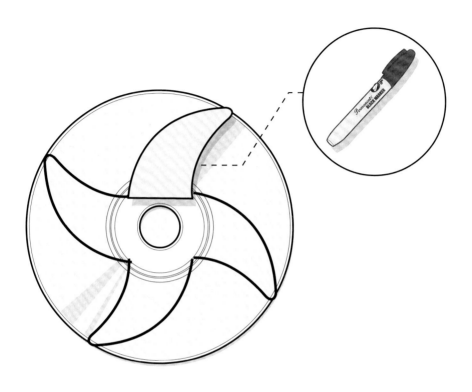

Place the sticky note pattern onto an unwanted CD or DVD, with the point extending to the edge of the disc. With the marker, trace the pattern four times, equally spaced, as shown.

# Step 3

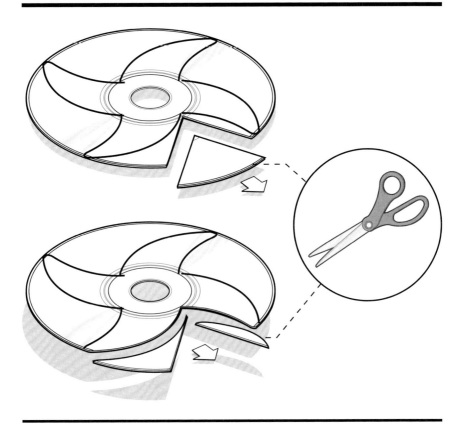

Cutting a CD or DVD with large scissors can be tricky because the disc might crack. To avoid this problem, slowly cut the disc apart in small sections. Start by removing basic straight wedges from the disc (top image). Next, clean up the blade edges with smaller curved cuts (bottom image). Repeat this step for all four sections until your disc resembles a throwing star.

# Step 4

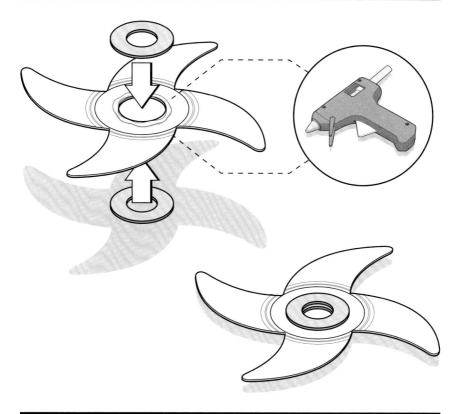

The throwing star is now functional. However, with a few additional steps you can increase its distance and durability.

To add weight to the throwing star, hot glue two metal washers or coins to the center of the disc, one on each side. Adding weight to both sides will add balance.

# Step 5

ELECTRICAL TAPE

Wrap the entire throwing star in electrical tape or duct tape. Not only will this reduce the possibility of the disc shattering on impact, but it will also eliminate any sharp edges left from the scissors.

The CD Manji Shuriken can be thrown horizontally or vertically using a motion similar to that for throwing a Frisbee. Hold the star parallel to your palm, gripping it between your thumb and index finger. Then use the weight of your body and arm in tandem to add power, letting the star slip from your fingertips. Be prepared: hold a stack of throwing stars in your opposite hand that can be slid off and flicked in rapid succession. Test your ninja star throwing skills with the Lantern Knockout (page 289).

***Remember to use eye protection!*** Throwing stars have pointed tips and can reach high speeds. ***MiniWeapon projects are not meant for use on living targets.*** Always stay clear of spectators and throw in a controlled manner.

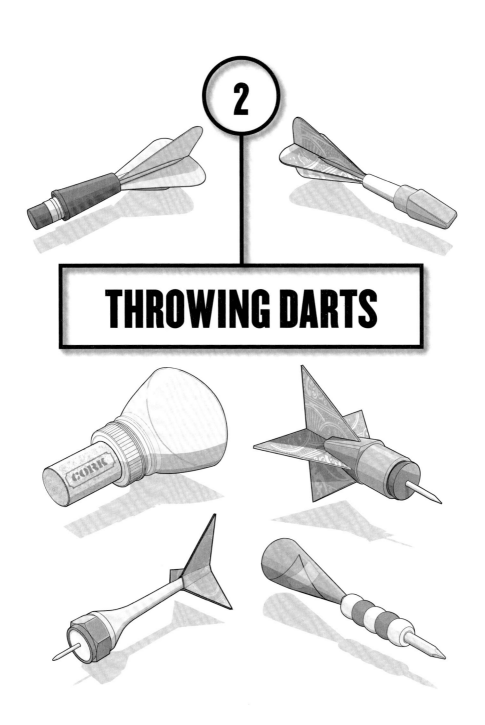

# THROWING DARTS

# BLUNT OFFICE DART

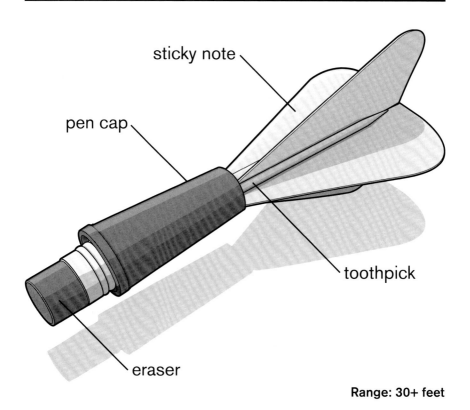

sticky note

pen cap

toothpick

eraser

**Range: 30+ feet**

The ninja master says, "He who dons the Blunt Office Dart will leave no path of destruction," because this dart is one of the most indoor-friendly ranged weapons in the MiniWeapons arsenal. Handcrafted with a soft eraser tip, molded plastic shaft, and sticky-note fins, it's slightly different from the traditional ninja one-piece metal *bo-shuriken* (steel spike) but can be utilized as a similar ranged weapon.

## Supplies

1 plastic ballpoint pen cap
1 wooden pencil
4 round wooden toothpicks
1 square sticky note (3 inches
    by 3 inches)

## Tools

Safety glasses
Large scissors
Needle-nose pliers
Hot glue gun

# Step 1

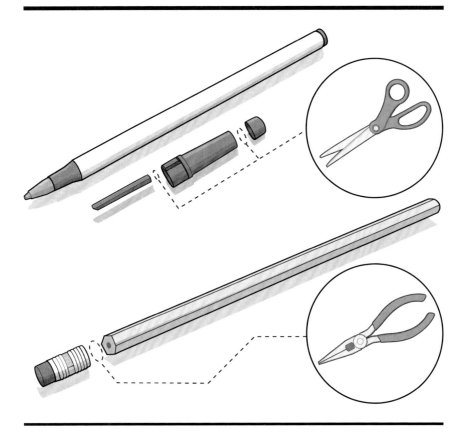

Start construction of the Office Blunt Dart by carefully using large scissors to remove the clip end and tip from a plastic ballpoint pen cap. After these parts are removed, the cap should be an open cylinder.

Next, use pliers to dislodge the metal eraser end from a wooden pencil. Try to avoid bending the metal flat when you pry it off.

# Step 2

Carefully hot glue the metal eraser assembly into the large opening in the pen cap, with the eraser facing out and protruding past the opening, as shown.

Next, assemble four round toothpicks into a square. Then hot glue the toothpick grouping into the removed end of the pen cap. These toothpicks will eventually hold the sticky note fins.

# Step 3

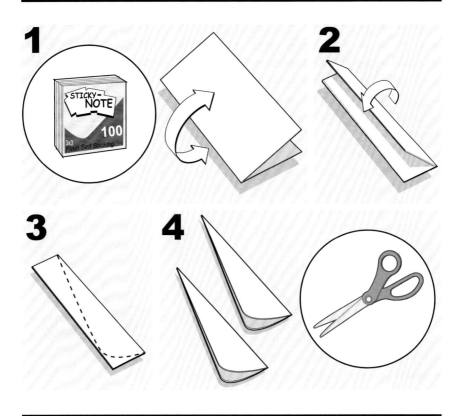

Now it's time to construct the dart's fins, also known as flights. First, fold a 3-inch-by-3-inch sticky note in half (1). Then divide it once more with an additional fold the long way, so that the final sticky note is ¼ the original width but still the original length (2).

As indicated by the dotted line, the center of the fin will be along the crease line. Drawing the fin pattern is optional (3). Use scissors to cut a triangular fin shape, with an optional round edge as shown (4). When you're finished, you should have two separate fins that are the same size.

# Step 4

Place the two sticky note fins side by side. On the center of the first fin, use scissors to cut a small slit from the top point of the triangle to about halfway down. On the center of the second fin, cut a small slit from the bottom edge to approximately halfway up the triangle. Now slide the two fins together to form the rear flight assembly.

# Step 5

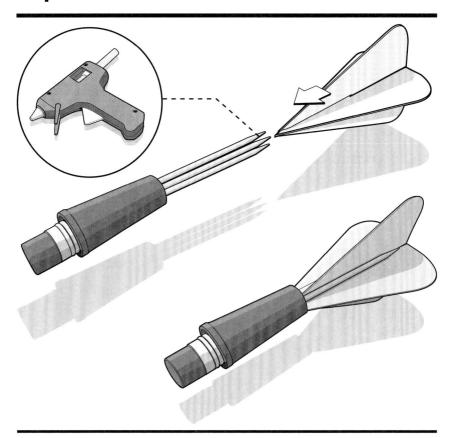

Now it's time to combine both dart assemblies. Dab a small amount of hot glue onto the tip of the fin assembly as indicated in the illustration. Then slide that assembly point-first between the fixed toothpicks, sandwiching each fin between two toothpicks.

Once the glue has dried, your dart is ready to throw. ***Even blunt darts are not meant for living targets.*** Always stay clear of spectators and throw darts in a controlled manner. Homemade weaponry can malfunction.

# PENCIL TOP ERASER DART

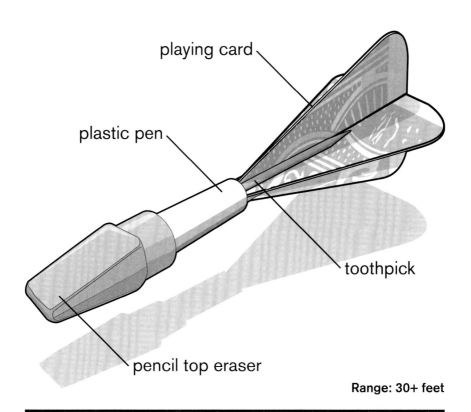

playing card

plastic pen

toothpick

pencil top eraser

**Range: 30+ feet**

The Pencil Top Eraser Dart makes a commanding sound when it bounces off its intended target! That sound is the result of a massive pencil top eraser tip, a high-impact plastic pen shaft, and durable coated cardstock fletching. This ninja MiniWeapon is built for rigorous training.

## Supplies

1 plastic ballpoint pen
1 pencil top eraser
4 round wooden toothpicks
1 playing card

## Tools

Safety glasses
Large scissors
Hot glue gun

# Step 1

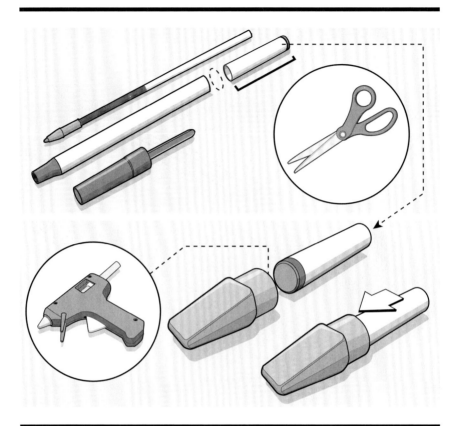

Disassemble one plastic ballpoint pen into its various parts. After removing the ink cartridge, carefully cut approximately 1¼ inches off the rear of the pen using large scissors.

Hot glue the pencil top eraser over the attached rear pen-housing cap. The enclosed rear housing cap will add some additional weight to the tip of the dart, which will increase accuracy.

# Step 2

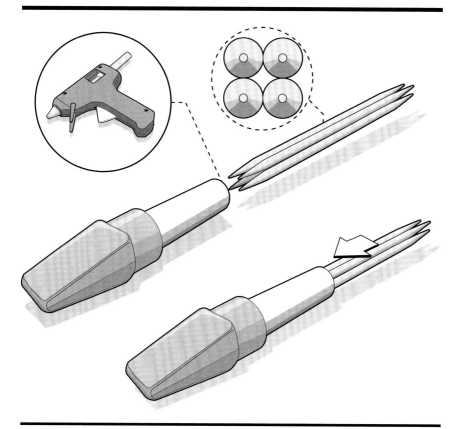

Add hot glue into the rear pen cylinder, then slide in four round tooth-picks in a square pattern, locking the grouping together with the hot glue. The toothpicks will hold the playing card fins.

# Step 3

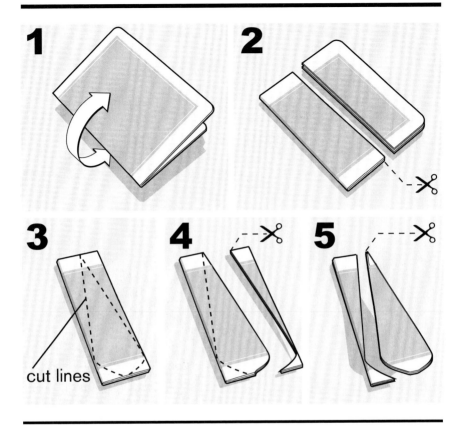

cut lines

Now it's time to construct the dart's fins, also known as flights. First, fold a playing card in half (1). With scissors, cut the folded card in half, with the crease along the height of the rectangle (2).

The fin pattern will be a tear-shaped diamond as illustrated by the dotted lines (3); drawing the pattern on the card is optional. With scissors, cut out one side of the fin design (4) and then repeat this cut on the opposite side so the fin is symmetrical (5).

# Step 4

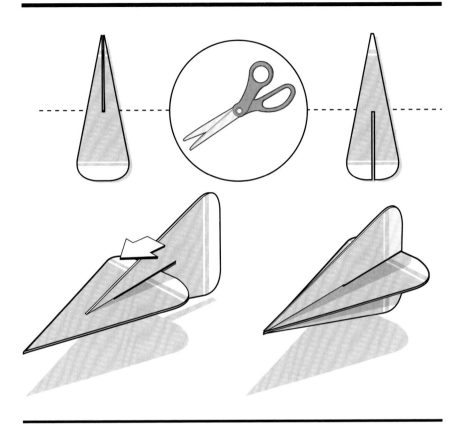

Place the two playing card fins side by side. In the center of the first fin, use scissors to cut a small slit from the top point of the fin to about halfway down. The width of the slit should be the same as the thickness of the card, but not bigger. In the center of the second fin, cut a small slit of the same width from the midpoint of the bottom edge to approximately halfway up the triangle. Now slide the two fins together to form the rear flight assembly.

# Step 5

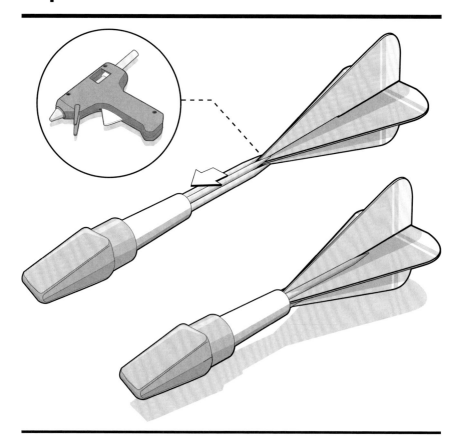

Now it's time to combine the dart and the fin assembly. Dab a small amount of hot glue onto the tip of the fin assembly as indicated in the illustration. Then slide that fin assembly point-first between the fixed toothpicks, sandwiching each fin between two toothpicks.

***Even blunt darts are not meant for living targets.*** Always stay clear of spectators and throw darts in a controlled manner. Homemade weaponry can malfunction.

# CORK BIRDIE

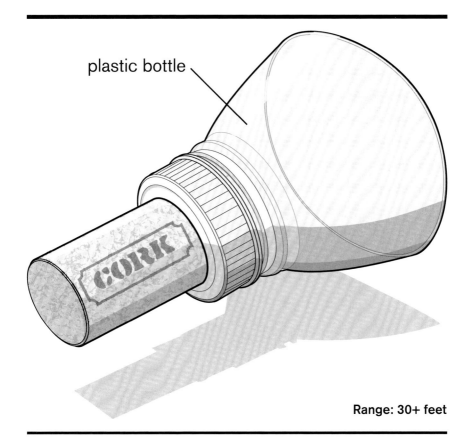

plastic bottle

**Range: 30+ feet**

The Cork Birdie, sometimes called a cork shuttle, cork dart, or cork shuttlecock, is an ingenious blunt-tipped throwing projectile with a unique conical tail. Thrown cork-first using *jiki da-ho* (the direct hit method), it doesn't require ninja super abilities to master, because it's quite aerodynamically stable. Because of its impressive range, the Cork Birdie can be utilized for both indoor and outdoor training exercises.

## Supplies

1 12-ounce plastic soft drink
  bottle, with cap
1 screw (½ inch to 1 inch long)
1 cork

## Tools

Safety glasses
Large scissors
Screwdriver

# Step 1

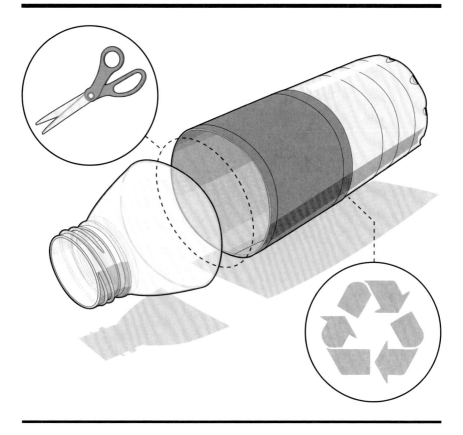

Using large scissors, cut off the top 2¼ inches from a 12-ounce plastic soft drink bottle. Recycle the bottom of the bottle.

# Step 2

Select a screw approximately 1 inch or less than the length of the selected cork. Position the screw tip in the center of the underside of the bottle's plastic cap and, with a screwdriver, screw it in until the screw head is flush.

Next, positioned at the center, hand screw the cork onto the fixed screw. You may need to use the screwdriver for a tight fit. The screw tip should **not** protrude from the cork's opposite end.

# Step 3

Almost finished! Just twist the finished cap assembly back onto the shortened bottle top. This conical tail makes the dart's flight extremely aerodynamic, which gives it a great range. Plus, with its soft cork tip, it's a great dart to toss around indoors. If you notice the cork breaking off or unscrewing, replace or adjust it.

***Even blunt darts are not meant for living targets.*** Always stay clear of spectators and throw darts in a controlled manner. Home-made weaponry can malfunction.

# PUSHPIN ERASER DART

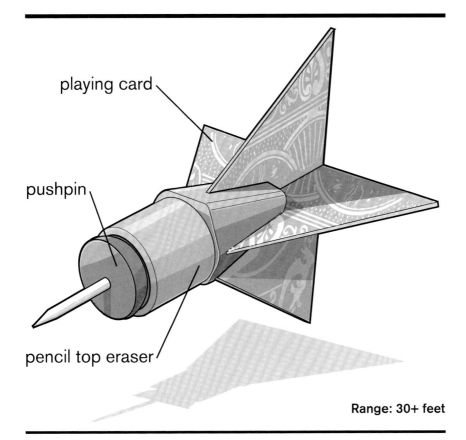

playing card

pushpin

pencil top eraser

**Range: 30+ feet**

With a heavy-gauge steel tip, the Pushpin Eraser Dart has all the elements of a menacing homemade dart. Master ninjas will appreciate this MiniWeapon's compact design and balanced weight distribution when target practicing.

## Supplies

1 pencil top eraser
1 pushpin
1 playing card

## Tools

Safety glasses
Hobby knife
Hot glue gun
Scissors

# Step 1

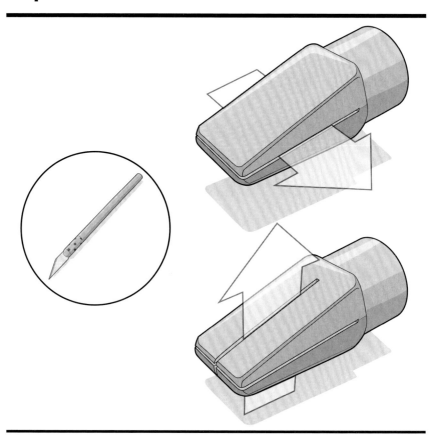

***Be especially careful with this step, which requires precision cuts using a sharp hobby knife; young ninjas should ask an adult for help.*** With the hobby knife, cut one ⅜-inch slit through the side of the pencil top eraser tip and then make an additional ⅜-inch slit through the center of the slope on the eraser tip. Both cuts should go completely through the eraser.

# Step 2

Dab a small amount of hot glue inside the eraser (or on the pushpin) and then slide the pushpin (point out) into the eraser opening, opposite the newly cut slits.

# Step 3

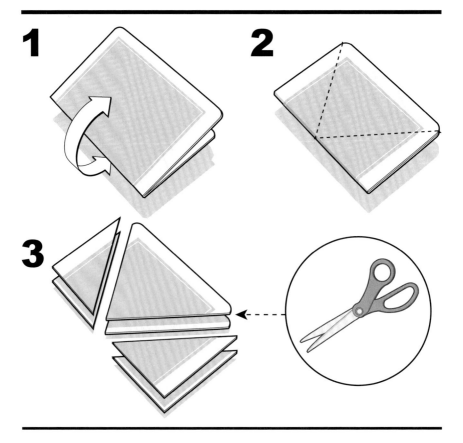

Now construct the dart's fins, also known as flights. First, fold one playing card in half (1). Then, as indicated with a dotted line (2), use scissors to cut out a triangle shape (3). When you're finished, you should have two separate triangles of the same size. Recycle the extra card material.

# Step 4

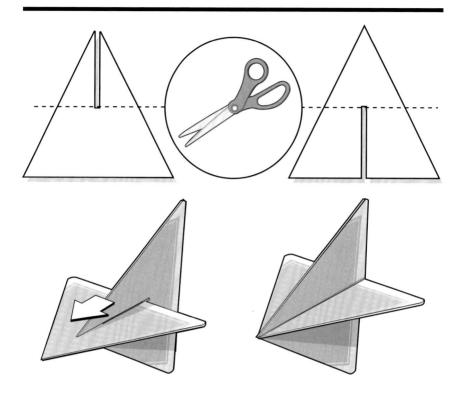

Place the two playing card triangles side by side. In the center of the first triangle, use scissors to cut a small slit from the top point of the triangle to about halfway down. The width of the slit should be the same as the thickness of the card, but not bigger. In the center of the second triangle, cut a small slit of the same width, but from the midpoint of the *bottom* edge to approximately halfway up the triangle. Now slide the two triangles together to form the rear fin assembly.

# Step 5

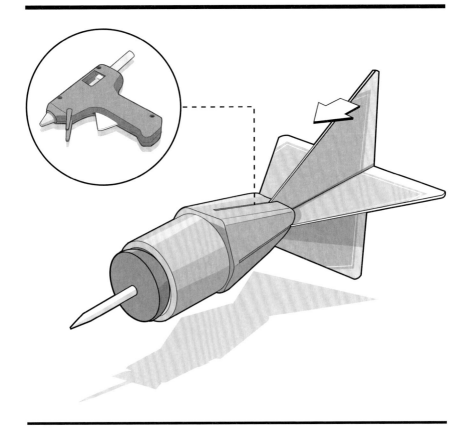

Now it's time to combine both the fin and the dart assembly. Dab a small amount of hot glue onto the tip of the paired fins as indicated in the illustration. Then slide that fin assembly into the eraser slits, aligning the fins with the four slots cut earlier.

Once the glue is dry, the Pushpin Eraser Dart is complete. We recommend target practice with the Chopstick Tripod (page 287).

***This dart is equipped with a dangerous point at the end and is not intended for living targets.*** Always stay clear of spectators and throw darts in a controlled manner. Homemade weaponry can malfunction.

# TEE THROWING DART

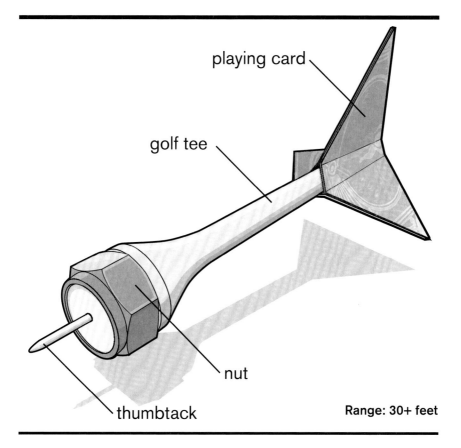

playing card

golf tee

nut

thumbtack

**Range: 30+ feet**

Traditional darts have been around for centuries. However, the ancient scrolls state, "If you don't leave the old path, how will you get on the new one?" The old ninja masters are right; the nontraditional design of the Tee Throwing Dart will definitely raise a few eyebrows around the dojo. It's constructed from a diverse bill of materials with one common goal: make one serious dart!

## Supplies

1 golf tee
1 metal nut (similar in diameter
    to golf tee head)
1 thumbtack
1 playing card

## Tools

Safety glasses
Hot glue gun
Scissors

# Step 1

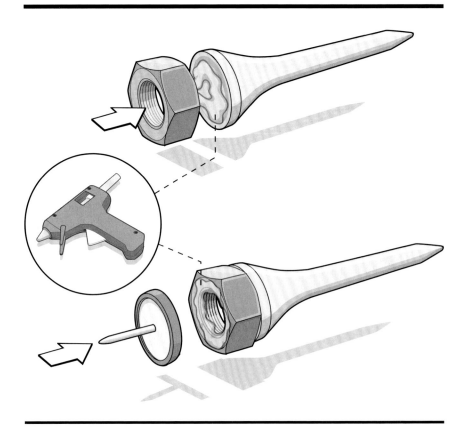

Locate a metal nut similar in diameter to the head of the selected golf tee (probably $5/16$ inch or $1/4$ inch). Hot glue the metal nut onto the golf tee head as illustrated and then hot glue a thumbtack, point out, on top of the fixed nut.

# Step 2

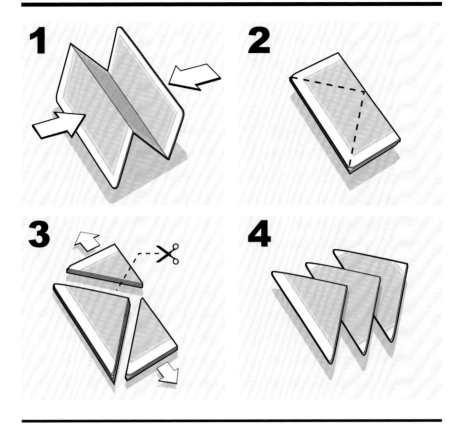

Accordion-fold one playing card with two crease lines, reducing the playing card to one-third its original width (1). The triangle fin pattern is represented by the dotted lines (2). With scissors, cut out the triangle fin, with the width of the original card as the base of the triangle (3). Once cut, separate the fins—an additional cut down the crease of one set will give you three identical fins (4).

# Step 3

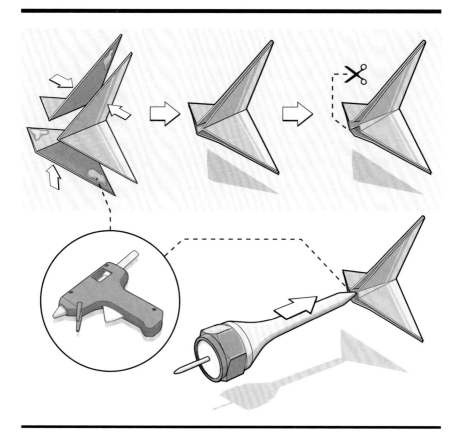

Crease all three playing card triangles down the center, then carefully hot glue them into a three-finned star, with each triangle point attached to the next. Test the fit by sliding the star onto the pointed tip of the golf tee. You may need to trim the tip of the assembly for a snug fit.

Next, dab a small amount of hot glue into the fin assembly center opening as indicated in the illustration. Then slide that fin assembly over the golf tee point. Once the glue is dry, the Tee Throwing Dart is complete. We recommend target practice with the Chopstick Tripod (page 287).

*This dart is equipped with a dangerous point at the end and is not intended for living targets.* Always stay clear of spectators and throw darts in a controlled manner. Homemade weaponry can malfunction.

# PONY BEAD NAIL DART

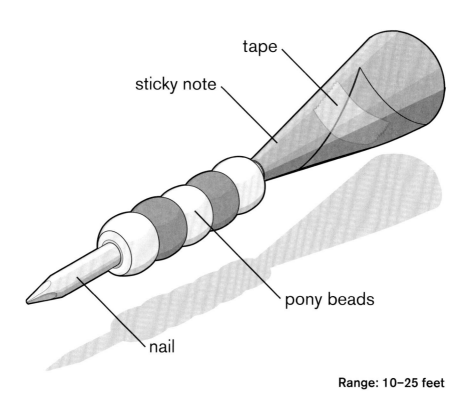

tape

sticky note

pony beads

nail

**Range: 10–25 feet**

With a solid galvanized nail core and reliable conical tail, the Pony Bead Nail Dart has proven reliable during critical missions. It has commanding aim and is built to stick into most intended targets. The narrow shaft allows color customization—just secure different-colored pony beads to distinguish between clans.

## Supplies

1 square sticky note (3 inches by 3 inches)
Clear tape
1 nail (approximately 2 inches long)
5 pony beads

## Tools

Safety glasses
Scissors
Hot glue gun

# Step 1

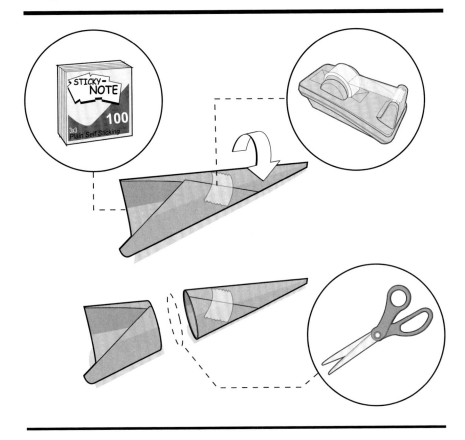

Start by making the dart conical fletch. Roll one 3-inch-by-3-inch sticky note or similar-sized paper into a funnel, as shown. Use a small amount of tape to hold the paper in place.

From the point, cut off approximately the last 2 inches of the cone for a clean diameter. This will help balance the dart.

# Step 2

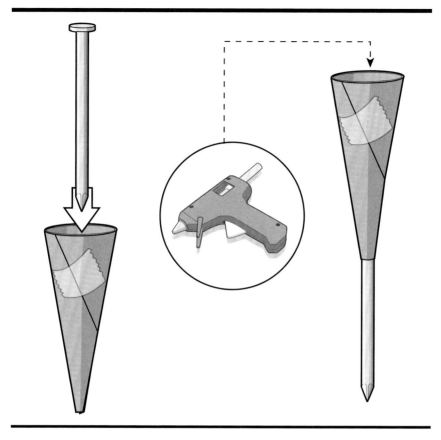

Slide one standard-headed galvanized nail—point first—into the open end of the sticky-note cone. Once in place, dab a small amount of hot glue into the cone to hold the nail in place.

# Step 3

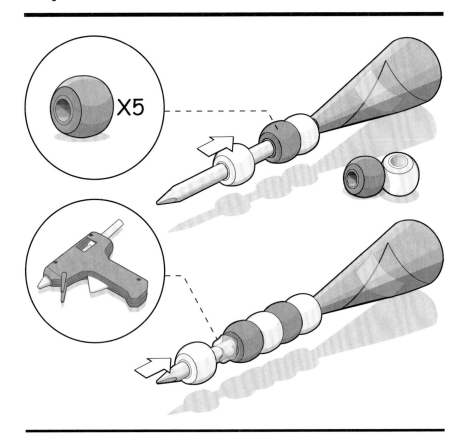

We recommend using multicolored pony beads for aesthetic reasons. Begin by sliding four beads onto the fixed nail as illustrated. With four beads against the conical fletch, add one more bead onto the nail and then dab hot glue between the fifth and fourth beads closest to the nail tip.

Once the glue is dry, the Pony Bead Nail Dart is complete and ready for ninja training. We recommend target practice with the Chopstick Tripod (page 287).

**This dart is equipped with a dangerous point at the end and is not intended for living targets.** Always stay clear of spectators and throw darts in a controlled manner. Homemade weaponry can malfunction.

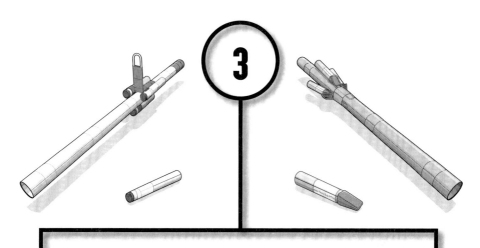

# BLUNT DART BLOWGUNS

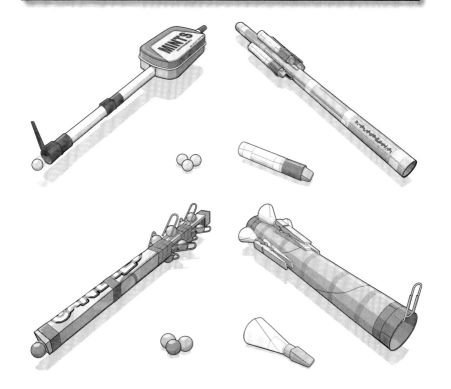

# BLUNT PEN BLOWGUN

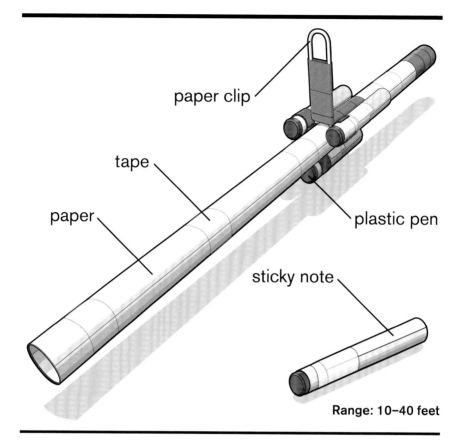

paper clip

tape

paper

plastic pen

sticky note

**Range: 10–40 feet**

Target practice without the risk of damaging the dojo with the Blunt Pen Blowgun! Crafted from one sheet of paper, modified sticky notes, and a few inexpensive pens, it's a quick build for a ninja in a time crunch. Plus, the blowpipe is cleverly designed to hold additional blunt darts for multiple strikes.

## Supplies

Clear tape
1 sheet of copy paper (8½ inches by 11 inches)
Electrical tape
1 large paper clip

## Tools

Safety glasses

Large scissors
Pliers

## Ammo

3+ plastic ballpoint pens with caps
1+ square sticky note (3 inches by 3 inches)

# Step 1

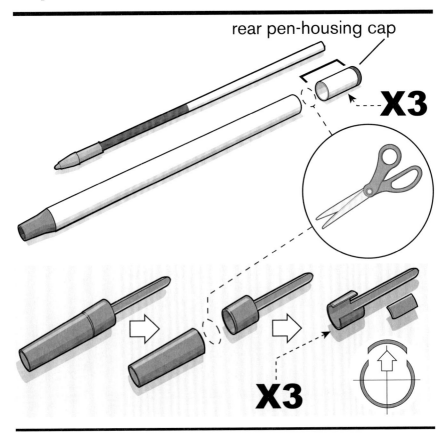

rear pen-housing cap

X3

X3

Remove the ink cartridges from three ballpoint pens. With large scissors, remove 1 inch from the back of each pen cylinder as shown, keeping the rear pen-housing cap in place.

Next, modify three pen caps to hold the blow darts. With scissors, cut each pen cap ¼ inch from the pen cap clip. Then remove a ⅜-inch section from the pen cap cylinder, centered opposite the pen clip. To test the dart clip, snap in the 1-inch pen cylinder; if the dart does not clip in, remove additional material from the pen cap as needed. Do not remove the attached pen clip.

# Step 2

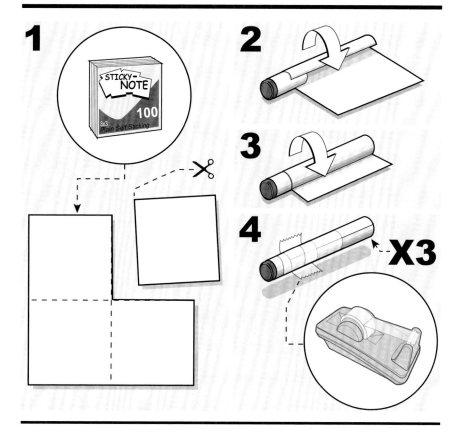

Each blow dart fletch will be constructed from a cut 3-inch-by-3-inch sticky note. Start by folding a sticky note in half. Then fold it one more time to create four stacked squares. Unfold the sticky note and use the 1½-inch crease lines as scissor guides to cut four 1½-inch paper squares (1).

Preroll the 1½-inch square prior to taping (2). Now, tightly roll the paper around the 1-inch pen cylinder, maintaining the same diameter as the pen cylinder (3).

Tape the rolled sticky note halfway up the pen cylinder with clear tape. Add additional clear tape to the middle of the sticky note to hold its form (4). The Blunt Pen Blowgun dart is complete. Create more darts with the remaining 1½-inch squares.

# Step 3

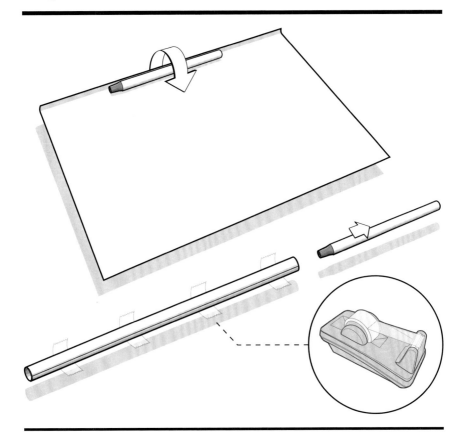

Using one of the leftover ballpoint pens as a diameter guide, roll the pen up (lengthwise) in a standard 8½-inch-by-11-inch sheet of paper. Prior to taping the newly formed cylinder, slightly unravel the sheet so the pen can easily slide out; it is important for the blow dart to have enough clearance to exit. Once you have increased the diameter of the newly formed paper tube, tape it in several places with clear tape.

# Step 4

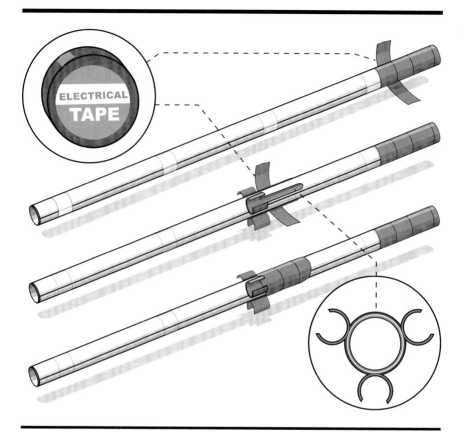

Wrap electrical tape around one end of the paper cylinder to create a protective mouthpiece. Without this added tape, the paper can quickly lose its cylindrical form due to dampness. Various types of tape will work, but the durability of electrical tape will help maintain the cylinder's form.

Next, approximately 5 inches from the taped mouthpiece, tape all three modified pen caps evenly spaced around the paper cylinder. Do not obstruct the custom clip detail. The final assembly should like the bottom image.

# Step 5

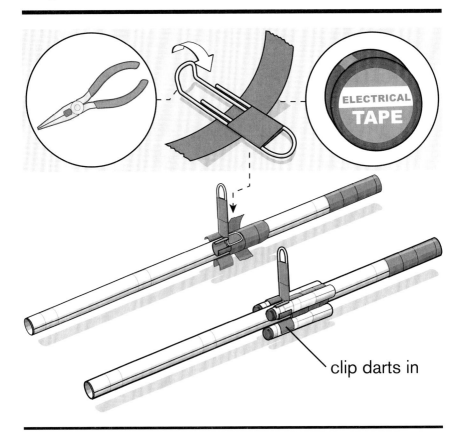

clip darts in

With pliers, bend a 90-degree angle approximately ¼ inch into a large paper clip. Then tightly tape around the paper clip's main body, but do not cover the top opening. Tape the small end of the paper clip in between the fixed pen cap clips on the blowgun. This will be the shooting sight. Just snap the darts into the clips and you are ready to go!

When you're ready to fire, load one pen dart into the paper tube with the blunt end facing toward the barrel exit. Take a deep breath, lift the blowgun to your lips, aim, and exhale sharply to blow the dart out.

*You should never inhale while the blowgun is in your mouth waiting to be launched.* Always be responsible and fire the blowgun in a controlled manner. ***Wearing eye protection is a must, as is staying clear of spectators.*** Firepower is limited by your respiratory muscles, and accuracy is limited by the blowgun's length and balance of the darts. Eventually the paper tube and homemade darts will lose their shape; replace them as needed.

# PENCIL TOP ERASER BLOWGUN

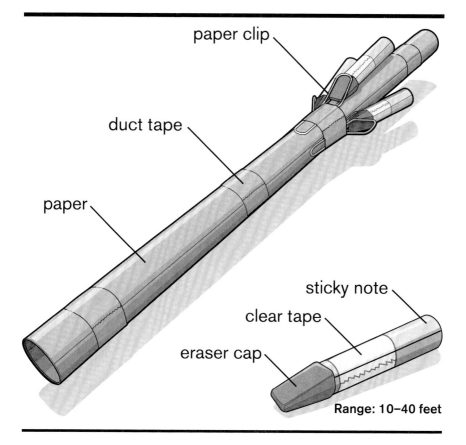

paper clip

duct tape

paper

sticky note

clear tape

eraser cap

**Range: 10–40 feet**

Sporting a duct tape–wrapped blowpipe and heavy eraser-tipped darts, the Pencil Top Eraser Blowgun has a commanding presence. Plus, the suggested blowpipe length can be doubled or even tripled depending on the user's range and accuracy preferences.

## Supplies

1 sheet of copy paper (8½ inches by 11 inches)
Duct tape
3 large paper clips
Clear tape

## Tools

Safety glasses

½-inch dowel (12 inches long)
Scissors
Hot glue gun
Needle-nose pliers

## Ammo

1+ square sticky note (3 inches by 3 inches)
3+ pencil top erasers

# Step 1

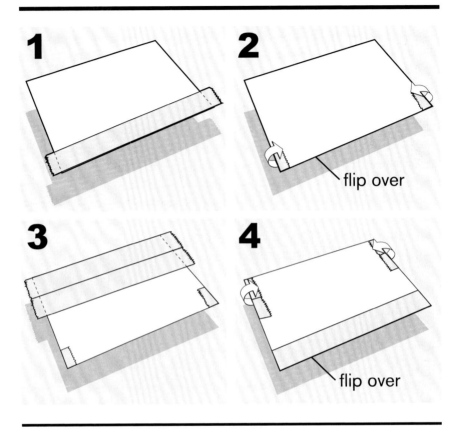

Start with an 8½-inch-by-11-inch sheet of paper. Place a single 2-inch-wide piece of duct tape along the long edge of the paper (1). Flip the paper over and fold over the excess tape on the back of the paper so it doesn't overhang (2).

Then, add two pieces of 2-inch duct tape along the opposite edge of the side with the folded-over tape tabs (3). Flip the paper over again to fold the tabs flush with the page (4).

# Step 2

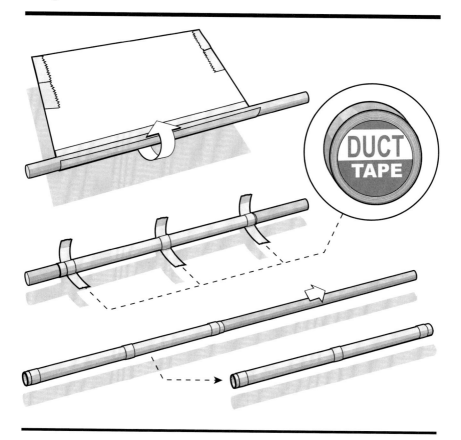

On a flat surface, lay a ½-inch dowel (or similar) along the top of the single strip of duct tape as illustrated (top image). Using the dowel as a guide, roll the paper into a tube. The finished tube should have a duct tape covering. Use multiple pieces of duct tape on the ends and center to hold the tube in place. When you're finished taping, slide the dowel out of the tube.

# Step 3

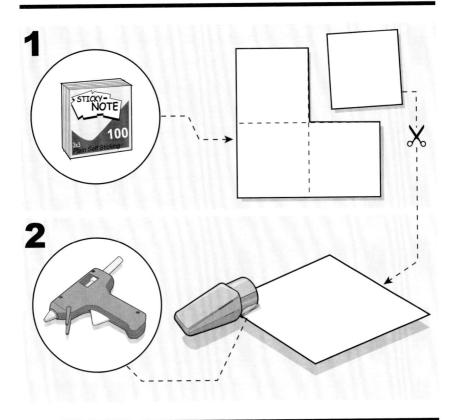

Each blow dart fletch will be constructed from a cut 3-inch-by-3-inch sticky note. Start by folding the sticky note in half, then fold it one more time to create four stacked squares. Unfold the sticky note and use the 1½-inch crease lines as guides. Use scissors to cut out four 1½-inch paper squares (1).

Preroll one 1½-inch paper square. Dab a small amount of hot glue onto the rear pencil top eraser fastener and then secure the end of the eraser onto the corner of the paper square, with the eraser angle hanging off the paper (2).

# Step 4

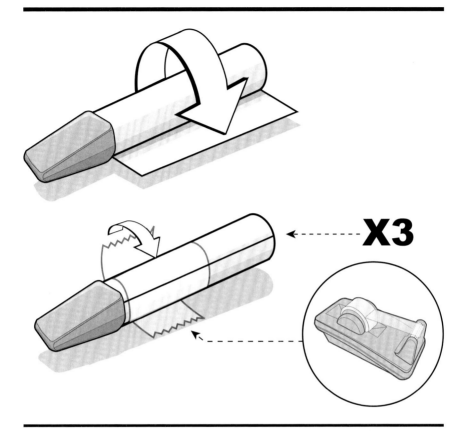

Now roll the attached paper around the eraser diameter. Use clear tape to secure the cylinder in place. We recommend making at least two additional darts by repeating steps 3 and 4.

# Step 5

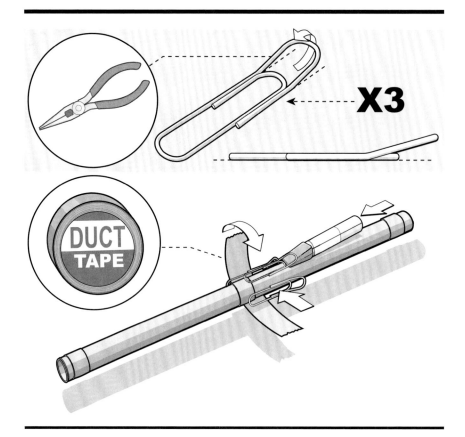

With pliers, ⅜ inches from one end, bend three large paper clips 20 degrees, as shown. Use duct tape to fasten the bent paper clips at even intervals around the blowpipe, with the 20-degree angle facing upward, approximately 4½ inches from the tube's entrance. These attached modified paper clips will hold the tips of the eraser darts. Test the clip angle by sliding the slope of the eraser under the 20-degree angle; the metal wedge should hold the angled dart. Adjust if needed.

When you're ready to fire, load one eraser dart into the paper tube with the blunt end facing the barrel exit. Take a deep breath, lift the blowgun to your lips, aim, and exhale sharply to blow the dart out. *You should never inhale while the blowgun is in your mouth.* Always be responsible and fire the blowgun in a controlled manner. *Wearing eye protection is a must, as is staying clear of spectators.* Firepower is limited by your respiratory muscles, and accuracy is limited by the blowgun's length and balance of the darts.

# MAGAZINE CAP BLOWGUN

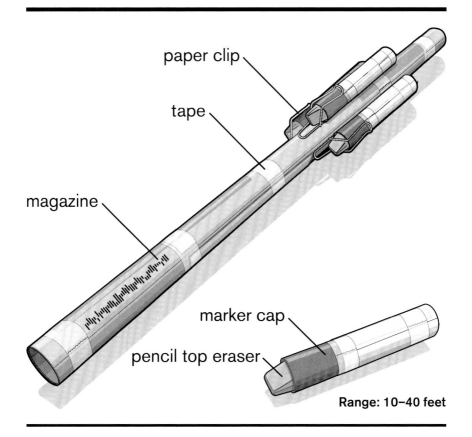

paper clip

tape

magazine

marker cap

pencil top eraser

**Range: 10–40 feet**

The Magazine Cap Blowgun is your subscription to mayhem. With one of the longer blowpipes in the MiniWeapon arsenal, not only will its size be noticeable, but its range will as well. In addition, the blowgun is equipped with practice-friendly soft-tip darts.

## Supplies

1 magazine
Clear tape
3 large paper clips

4 or 5 large art markers, or 1
  ½-inch dowel
Hot glue gun
Needle-nose pliers

## Tools

Safety glasses
Staple remover
Scissors
Ruler

## Ammo

1+ standard art marker cap
1+ pencil top eraser
1+ magazine page

# Step 1

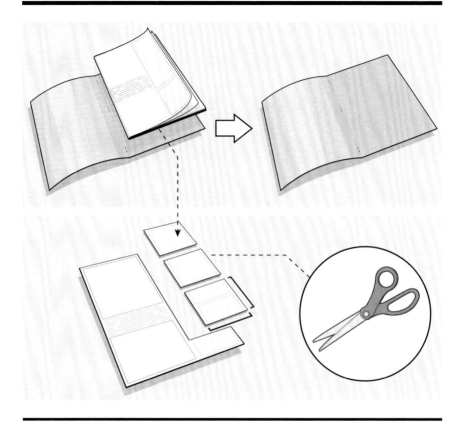

The blowpipe will be constructed from the cover of a standard magazine. Carefully remove the staples from the magazine cover, avoiding unnecessary rips.

The conical fletch of the blow dart will be created from one of the removed pages of the magazine. With scissors and ruler, measure and cut at least three 2½-inch-by-2½-inch squares from one of the pages.

# Step 2

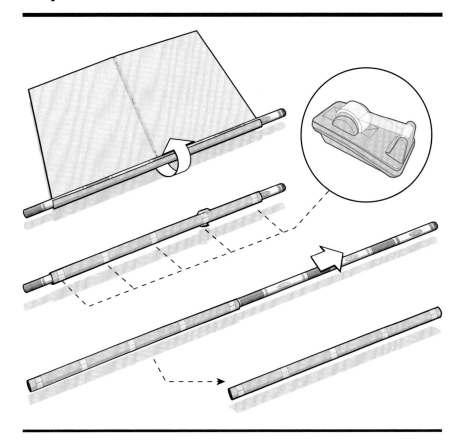

To start construction, create a blowpipe diameter guide out of four or five large art markers snapped together, or a ½-inch dowel, with a total length greater than the open magazine.

On a flat surface, lay the marker assembly along one edge of the magazine cover length. Using the marker assembly, slowly roll up the cover until you have created a tight and even roll, then secure the roll using several pieces of tape at the ends and middle. Once the tape is in place, slide the markers out of the tube to finish the blow tube.

# Step 3

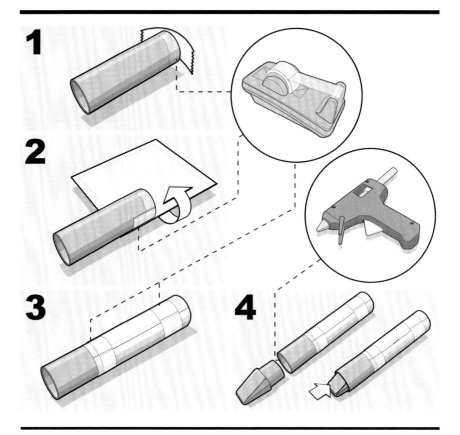

The main body of the blow dart will be constructed from a large marker cap. To ensure that the dart is airtight, add a small piece of tape over the marker end of the cap (1). Then tape half the cap onto the corner of the 2½-inch-by-2½-inch magazine square, with the airtight taped end facing the magazine square (2).

Roll the magazine section around the attached marker cap, maintaining a similar diameter, and then add tape to the front and rear of the magazine cylinder to hold its form (3). For a soft-tip effect, hot glue a pencil top eraser to the top of the attached cap. Some caps may not require glue because of a snug fit (4).

# Step 4

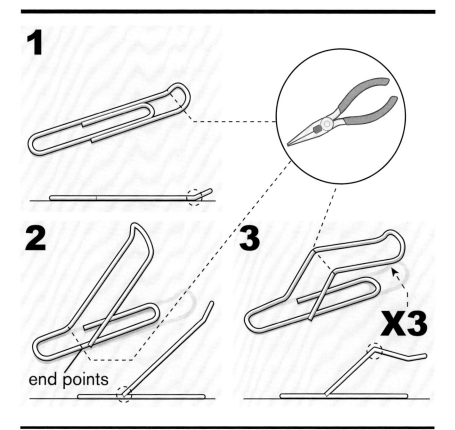

**1**

**2**

end points

**3**

**X3**

The blow dart holders will be created from three modified large paper clips. Use pliers at the end of the paper clip with two hoops to bend one 45-degree angle ⅛ inch past the first hoop (1).

Next, with pliers, bend the paper clip upward at a 45-degree angle at the paper clip end points (2).

At the halfway point of the 45-degree angle, bend the paper clip downward at negative 10 degrees (3) using pliers. This last bend will create a hook to catch and hold the blow darts.

# Step 5

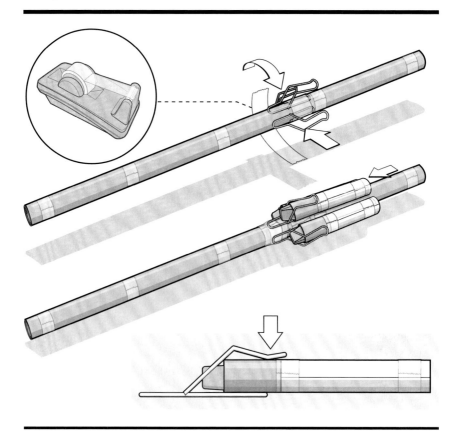

Evenly spaced around the blowpipe's diameter, tape all three modified paper clips approximately 7 inches from the entrance of the blowpipe, with the clip detail facing the closest entrance. Slide all three blow dart eraser ends into the clips as illustrated, with the eraser extending through the angled paper clip as a second hold point. Adjust the paper clip angles if needed.

When you're ready to fire, load one eraser dart into the paper tube with the blunt end facing toward the barrel exit. Take a deep breath, lift the blowgun to your lips, aim, and exhale sharply to blow the dart out. *You should never inhale while the blowgun is in your mouth waiting to be launched.* Always be responsible and fire the blowgun in a controlled manner. *Wearing eye protection is a must, as is staying clear of spectators.* Firepower is limited by your respiratory muscles, and accuracy is limited by the blowgun's length and balance of the darts.

# AIRSOFT TIN BLOWGUN

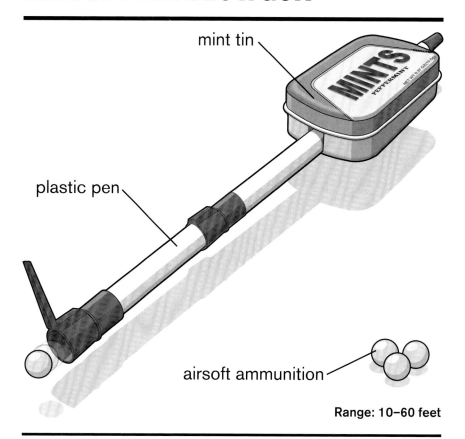

mint tin

plastic pen

airsoft ammunition

**Range: 10–60 feet**

The ninja master says, "Speak softly but carry an accurate blowgun." The Airsoft Tin Blowgun fits the bill. Its precision targeting starts with its high-impact plastic blowpipe, which can be lengthened to increase its range or disassembled for concealment. In addition, its frame boasts ammo storage that clamps shut when you're on the move!

## Supplies

2 plastic ballpoint pens with caps
1 small mint tin (1½ inches by 2¼ inches, or 0.37 ounces [10.5 grams])

## Tools

Safety glasses
Large scissors
Hot glue gun
Wire cutters or pliers

## Ammo

1+ airsoft BBs (0.12 grams)

# Step 1

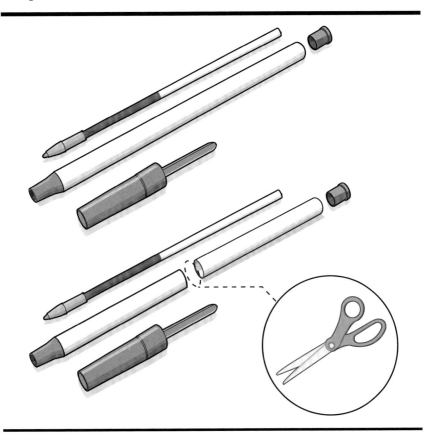

Start with two plastic ballpoint pens, with the same dimension and diameter. Disassemble each pen into its various parts. Depending on the pens, you may need a tool to dislodge the rear pen-housing caps. (A hobby knife or small pliers should be sufficient.) Save the rear housing caps for building the Blunt Pen Blowgun (page 77); the ink cartridges can be discarded. Keep the plastic pen tip connected to one of the pen housings; this is the safety mouthpiece.

Optionally, use large scissors to cut one of the pen housings in half, which will reduce the blowpipe barrel to a more manageable, pocket-length size. However, the smaller size may reduce the blowgun's range and accuracy. If you do not cut the pen housing, remove the plastic pen tip so the BB is unobstructed when it passes through the pen shaft.

# Step 2

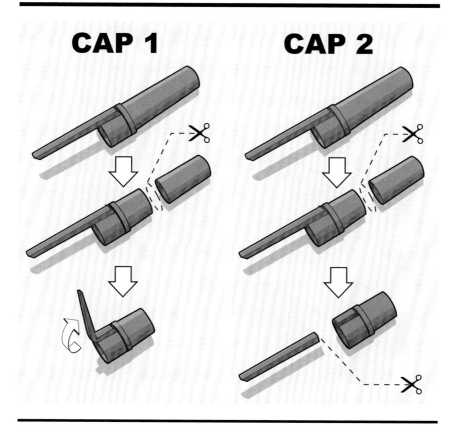

**CAP 1**  **CAP 2**

Both pen caps will need to be modified to assemble the pen blowpipe.

Starting with cap 1 (left illustration), cut the cap in half using a pair of large scissors. Finish modifying the cap by bending the pen clip upward at a 90-degree angle. Discard the cut pen cap tip. This 90-degree clip detail will be the blowgun's targeting sight.

The second cap should also be cut in half. Also remove the pen clip.

# Step 3

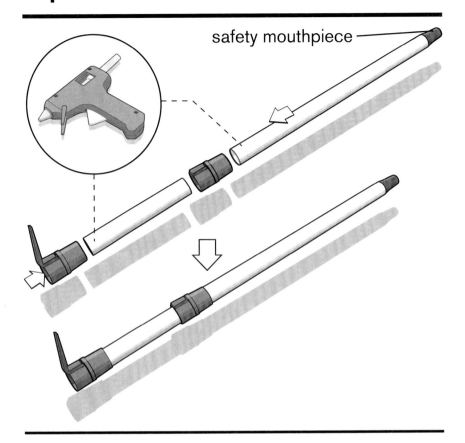

safety mouthpiece

Start the blowgun assembly by hot gluing the pen cap with 90-degree pen clip to the end of the halved pen shaft without the safety mouthpiece (or full housing, depending on your preference in step 1). Hot glue the second halved pen cap onto the full pen shaft, opposite the safety mouthpiece. Do not allow hot glue to enter the pen shaft or it could obstruct the ammo.

Test fit the two assemblies. If the first pen shaft fits snugly into the second pen shaft, there is no need to add hot glue, giving you the option to disassemble the blowgun for concealment. If the connection is loose, hot glue the assembly together, being careful not to get hot glue inside the pen shaft.

# Step 4

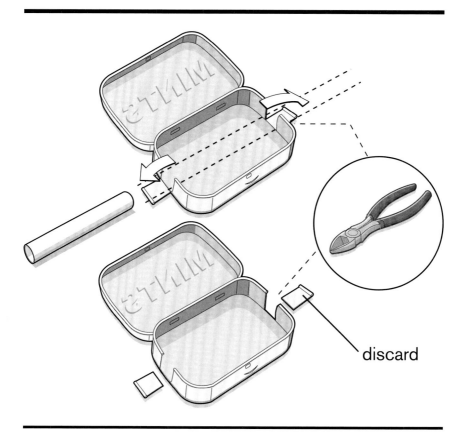

discard

This blowgun will have an integrated ammo clip, which is designed from a modified small mint tin. Large tins or plastic containers can be substituted.

With the tin container open, make two cuts equal to the pen shaft diameter in the center of both ends of the tin walls (top image) using wire cutters or pliers that are capable of snipping tin.

Fold down the tin flaps and remove both flaps with the cutters. Discard the flaps and make sure the attached tin cover can still close. Restore the metal walls of the container if they were bent during the removal of the flaps.

# Step 5

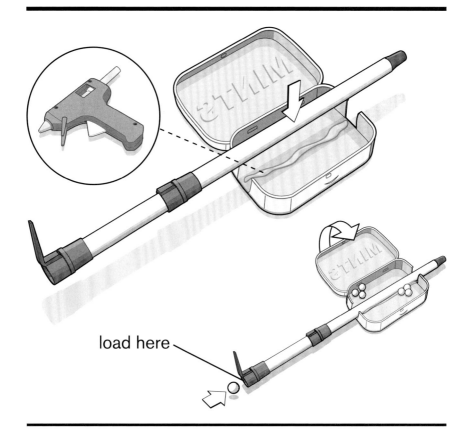

load here

Almost complete! Prior to hot gluing, place the pen shaft inside the tin, to see if the cover can close. Remove additional tin if the top is ajar.

With the barrel 90-degree sight upward, hot glue the pen blow-pipe inside of the tin, snug between the walls, with the safety mouthpiece protruding about 1 inch. Allow the glue to dry prior to loading.

The safety mouthpiece will prevent the user from accidently inhaling the plastic BBs. So with the safety mouthpiece fixed to the blowpipe shaft, drop one BB down the blowpipe exit and let it roll to the back. *If it falls through the safety mouthpiece, do not use this device—the opening is too large.* Take a deep breath, lift the blowgun to your lips, aim, and exhale sharply to blow the BB out.

*Never inhale while the blowgun is in your mouth. Eye protection is a must, as is staying clear of spectators.* Firepower is limited by your respiratory muscles, and accuracy by the blowgun's length. *Remember that BBs can ricochet on impact.*

# CANDY BOX BLOWGUN

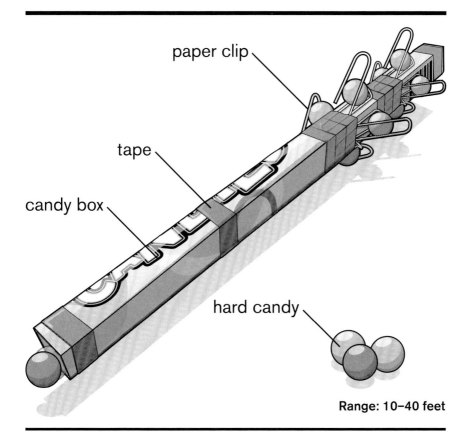

paper clip

tape

candy box

hard candy

**Range: 10–40 feet**

The Candy Box Blowgun is a perfect example of a ninja's ability to transform everyday objects into contraptions of trouble! This sweet design turns a large box of candy into a threatening MiniWeapon that is built to launch round, hard candies at unbelievable distances. The out-of-box thinking doesn't stop there—attached to the blowpipe are modified paper clips that hold additional rounds of ammo.

## Supplies

1 small candy box (6 ounces)
Electrical tape
8 large paper clips

## Tools

Safety glasses
Pen

Ruler
Scissors
Pliers

## Ammo

1+ round candies

# Step 1

Start with an empty 6-ounce candy box or small cardboard box similar in size. Disassemble the box by carefully peeling apart the factory-glued tabs that hold the box together. The finished package should lie flat and completely open, with the graphics on the bottom.

# Step 2

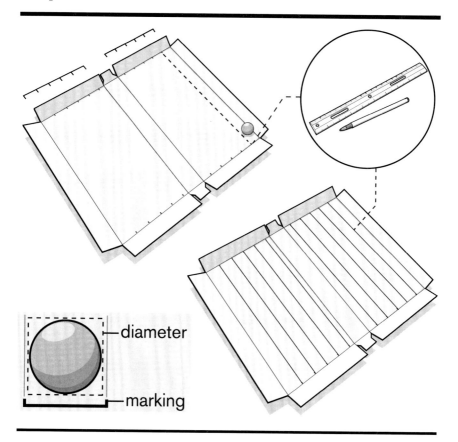

diameter

marking

The sizes of hard round candies vary, so use a ruler to measure the diameter of your selected candy, then increase the measurement slightly to avoid blowpipe jams. For example, Gobstoppers have a rough diameter of ½ inch. So you would add ⅛ inch to the ½ inch for a final measurement of ⅝ inches.

With a pen and a ruler, make a series of 12 marks at ⅝-inch intervals (or whatever your final measurement is) along the top and bottom box panel crease lines, on both original height axes. Use a ruler to connect the lines to create five ⅝-inch columns on each top and bottom panel, which will eventually become the walls of the blowpipe.

# Step 3

With scissors, cut out the two box panels with guidelines. Each removed panel should have five drawn columns. Do not cut down the drawn lines.

# Step 4

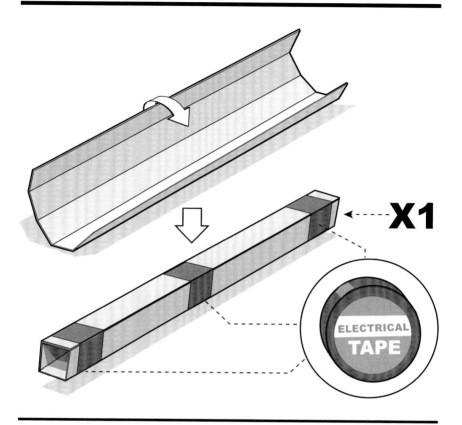

**X1**

ELECTRICAL
TAPE

Crease each drawn line inward at a 90-degree angle to form a four-sided box, with the fifth panel overlapping the first for added strength (top image). Once folded into a box, tape the square tube with several pieces of electrical tape, one piece at each end and one piece in the middle (bottom image).

At this point, roll a candy through the tube and test for obstructions. If the ammo becomes lodged in the blowpipe, refold it to make it slightly larger, and adjust the fold lines on the second cardboard rectangle.

# Step 5

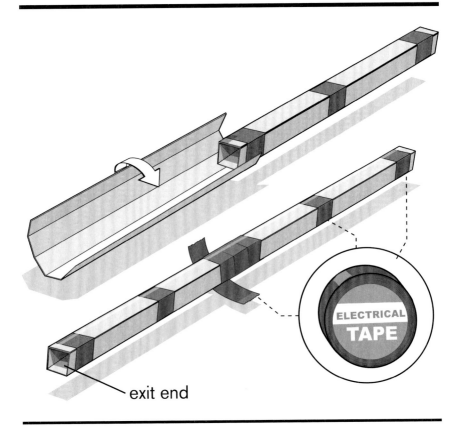

exit end

Prefold the second cardboard rectangle like you did the first, using the pen lines as a guide and folding each panel inward at a 90-degree angle to form a square.

Extend the blowpipe by attaching the second rectangle ¾ inch onto the previously constructed tube. Once in place, fold the cardboard around the tube, with the fifth panel overlapping the first for added strength. Add tape to the ends and center of the tube to hold it in place, using added tape to secure both tubes. The newly added tube will also be the end of the blowpipe's exit.

# Step 6

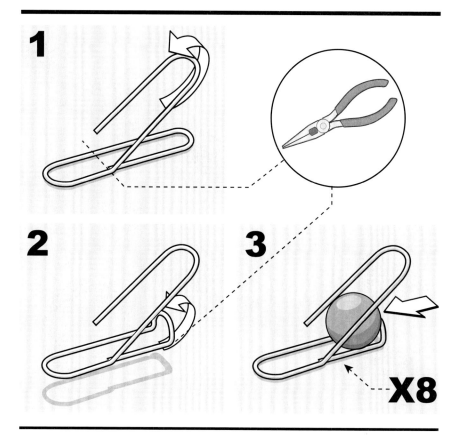

The optional candy ammo clips will be created from eight large modi-fied paper clips; however, four ammo clips are also sufficient.

Use pliers to bend the large loop upward at a 45-degree angle (1) at the paper clip double-loop end points. Then, approximately ¼ inch under the 45-degree loop, bend the small loop 90 degrees upward (2).

Test the fit by snapping one round candy into the clip (3). The candy should lock behind the small 90-degree angle, held in by the pressure of the 45-degree angle. If the candy falls out, reduce the angle.

# Step 7

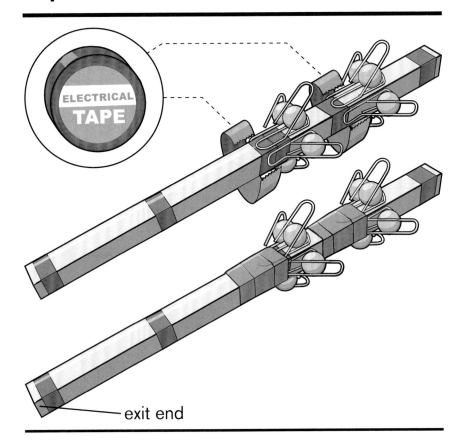

exit end

With electrical tape, tape two sets of four ammo clips on the barrel. The first set will be placed approximately 5 inches from the entrance. Use tape over the flat ends of the paper clips to hold the set in place. The second set of clips will be attached approximately ½ inch back from the first. Snap the round candies into the clip and this blowgun is mission ready! Note: this blowgun is designed to shoot candies; however, a safety mouthpiece can be constructed from an additional modified paper clip, similar to the safety mouthpiece on the Playing Card Blowgun (page 141).

Load one round candy into the blowpipe entrance. Take a deep breath, lift the blowgun to your lips, aim, and exhale sharply to blow the candy out. *Never inhale while the blowgun is in your mouth. Eye protection is a must, as is staying clear of spectators.* Firepower is limited by your respiratory muscles, and accuracy by the blowgun's length. *Remember that hard candies can ricochet on impact.*

# CARDBOARD TUBE BLOWGUN

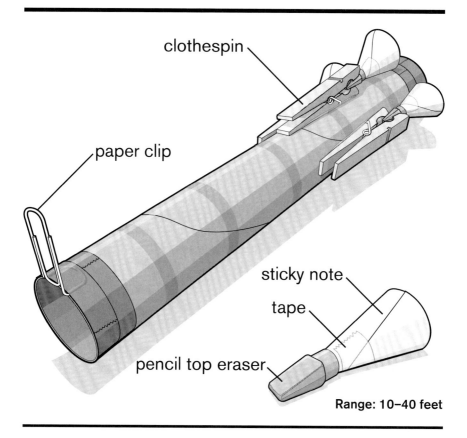

clothespin

paper clip

sticky note

tape

pencil top eraser

**Range: 10–40 feet**

With a solid blowpipe and a heavy dart design, the Cardboard Tube Blowgun is built for harsh training. The blowpipe takes only seconds to prep, giving you plenty of time to assemble additional darts.

## Supplies

1 cardboard tube (1-inch diameter or less)
4 wooden clothespins
1 large paper clip
Electrical tape
Clear tape

## Tools

Safety glasses
Hot glue gun
Pliers
Scissors

## Ammo

1+ square sticky note (3 inches by 3 inches)
1+ pencil top eraser

# Step 1

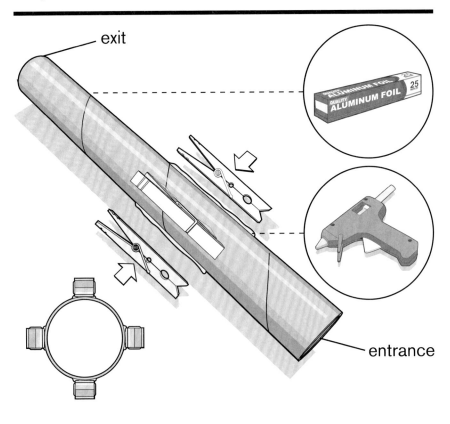

Select a small-diameter cardboard tube to construct the blowgun housing. An aluminum foil or plastic-wrap tube will work best; most of these tubes have a diameter of less than 1 inch. A tube with a diameter larger than 1 inch will dramatically decrease the firing distance of the eraser darts.

Hot glue four wooden clothespins evenly spaced around the middle of the cardboard tube, with the clip end 3½ inches from the entrance.

# Step 2

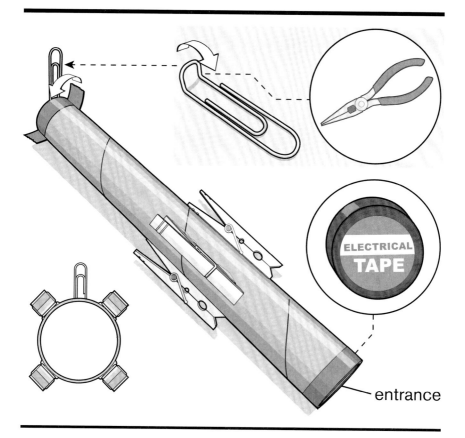

entrance

To create an optional barrel sight, use pliers to bend one loop of a large paper clip to a 90-degree angle approximately ¼ inch from one end of the clip. With electrical tape, attach the small end of the paper clip to the tube at the end opposite the clothespins, fixed between the clothespins to avoid obstructing the user's view.

Wrap electrical tape around the entrance end of the cardboard cylinder to create a protective mouthpiece. Without this tape, over time the tube will lose its form due to dampness.

# Step 3

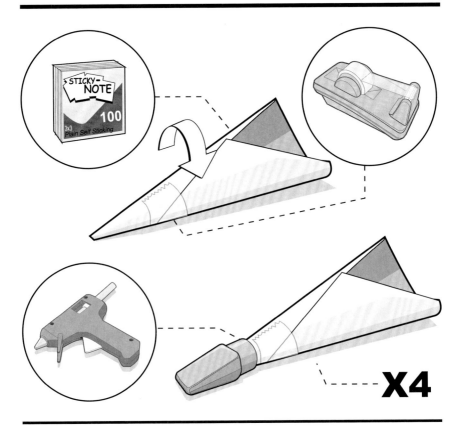

To make the conical fletch, roll one 3-inch-by-3-inch sticky note into a funnel as shown. Use a small amount of clear tape to hold the paper cone in place.

Next, dab a small amount of hot glue inside the eraser opening and then slide the paper fletch into the eraser opening. Repeat these steps until you have completed the suggested four eraser darts.

# Step 4

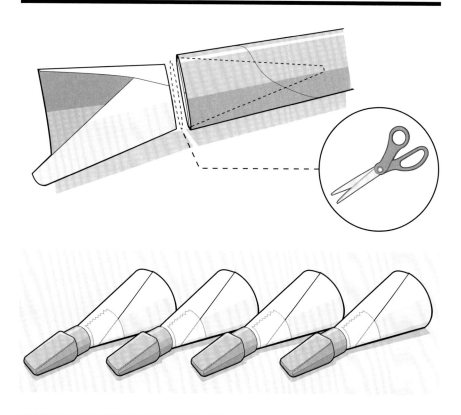

Next, insert one of the constructed darts into the cardboard tube, eraser tip first. Once the funnel is gently wedged in the tube, use scissors to trim off the extra cone material hanging out of the tube. The blow dart's maximum width is now the same diameter as the blowgun, which will increase the contact surface when you exhale and launch the dart through the gun. Repeat the trimming on the remaining three eraser darts.

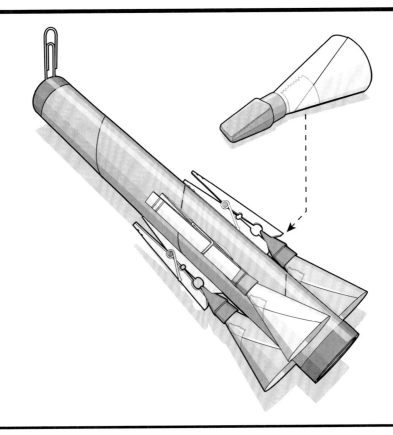

The Cardboard Tube Blowgun is done! Just clip the eraser ends into the clothespins around the blow tube as shown.

To fire, load one of the eraser darts into the entrance of the blowgun, eraser first. Take a deep breath, lift the blowgun to your lips, aim, and exhale sharply to blow the dart out.

It is very important to remember that you're firing darts from your mouth. *You should never inhale while the blowgun is in your mouth waiting to be launched.* Always be responsible and fire the blowgun in a controlled manner. *Wearing eye protection is a must, as is staying clear of spectators.* Firepower is limited by your respiratory muscles, and accuracy is limited by the blowgun's length.

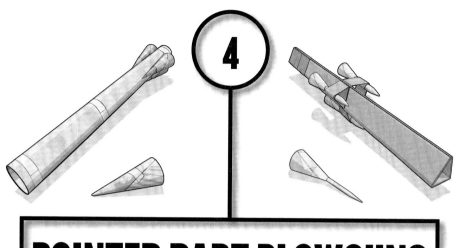

# POINTED DART BLOWGUNS

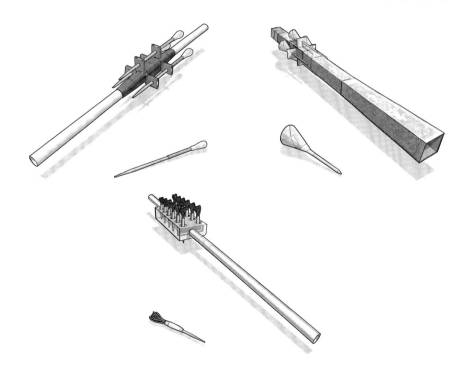

# NEWSPAPER BLOWGUN

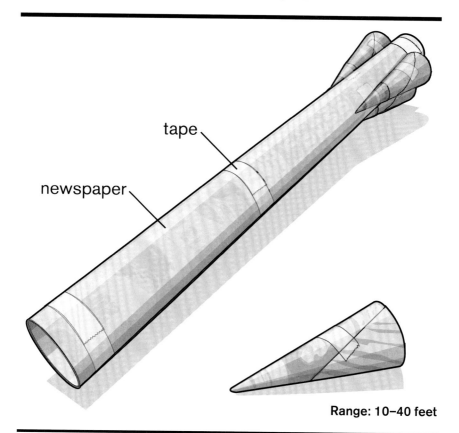

newspaper

tape

**Range: 10–40 feet**

Extra! Extra! Read all about it! The Newspaper Blowgun surprises skeptics! The headlines don't lie. With a possible range of 40 feet, the Newspaper Blowgun will convince many doubters. Built from little more than a single newspaper, it's inexpensive and accurate—exactly what yen-pinching ninjas love!

## Supplies

1 newspaper
Clear tape

## Tools

Safety glasses
Scissors
Ruler (optional)

## Ammo

1+ piece of newspaper (3 inches by 3 inches)
1+ paper clip (optional)

# Step 1

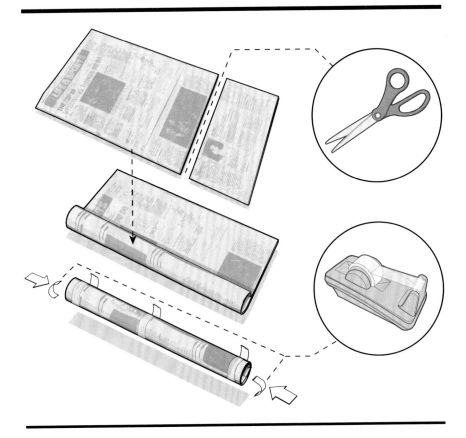

Remove a single newspaper spread and position the crease vertically, with the crease along the left side. With scissors, cut off the bottom 3 inches of the newspaper. (The removed material will be used later to create the blow darts.)

Roll up the main newspaper into a tube with a rough diameter of 1 inch. Use several piece of tape to secure the tube–at least two at the ends and one in the center of the tube. The blowgun is complete.

# Step 2

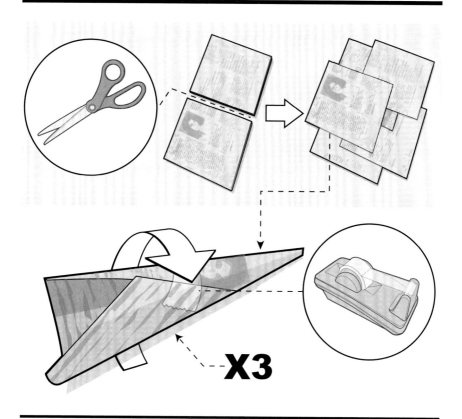

Now construct the blow darts. With scissors, cut the leftover newspaper into three or four 3-inch-by-3-inch squares.

Next, take one 3-inch-by-3-inch paper section and roll it into a funnel. Use tape to hold the paper in place and then make at least three more funnels with the remaining paper.

# Step 3

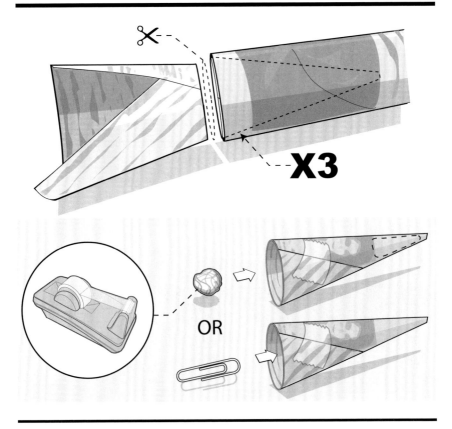

With the point first, insert one of the paper funnels into the newspaper tube. Once the funnel is gently wedged in the tube, use scissors to trim off the extra cone material hanging out of the tube. The blow dart's maximum width is now the same diameter as the blow tube, which will increase the contact surface when you exhale and launch the dart through the tube. Now trim the remaining paper funnels.

To increase the darts' shooting distance, you can wedge a tape ball or paper clip into the point of the cone. This also increases the tip weight for added accuracy.

# Step 4

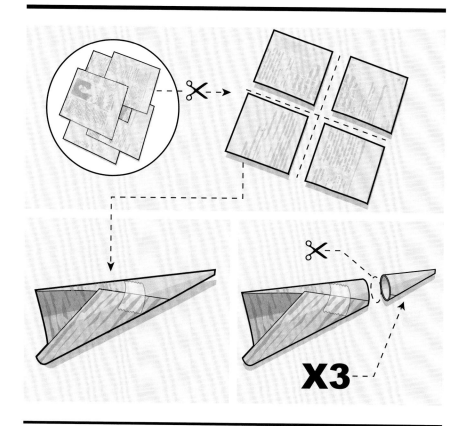

Optional dart holders can be crafted from newspaper scrap and attached to the blowpipe. To create holders, start with one 3-inch-by-3-inch newspaper square, then divide and cut that square into four equal sections (top).

Next, roll one paper section into a funnel. Use tape to hold the paper funnel in place and add additional tape to the tip of the funnel. Then, with scissors, reduce the funnel's total length by 1 inch. Create three more funnels similar in size following these steps.

# Step 5

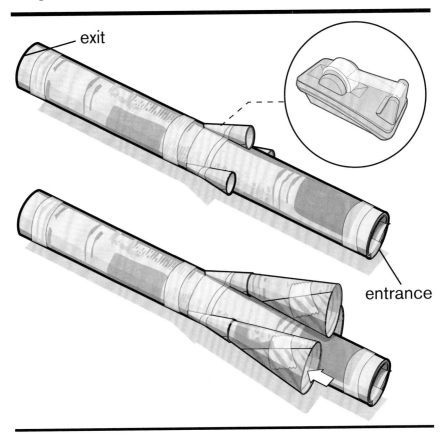

exit

entrance

With tape, secure all three dart holders (small cones) evenly spaced around the blowpipe, with the cone openings approximately 5 inches from the entrance, as shown. Once the cones are attached, they can carry the newspaper darts by wedging the tips into the holders. The pressure should hold them in place.

To fire, load one of the weighted newspaper darts into the entrance of the blowpipe, point first. Take a deep breath, lift the blowgun to your lips, aim, and exhale sharply to blow the dart out.

It is very important to remember that you're firing darts from your mouth. *You should never inhale while the blowgun is in your mouth waiting to be launched.* Always be responsible and fire the blowgun in a controlled manner. *Wearing eye protection is a must, as is staying clear of spectators.* Firepower is limited by your respiratory muscles, and accuracy is limited by the blowgun's length.

Pointed Dart Blowguns

# CEREAL BOX BLOWGUN

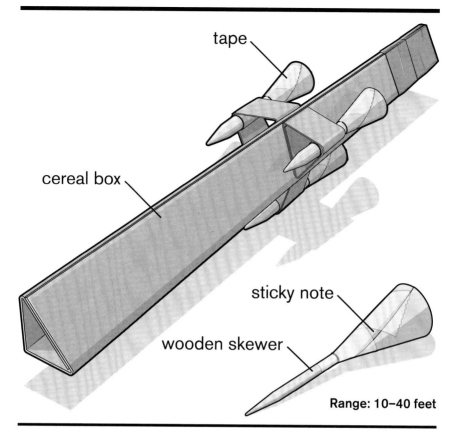

tape

cereal box

sticky note

wooden skewer

Range: 10–40 feet

Ninja wisdom reminds us to look beyond the obvious. This repurposed cereal-box-turned-blowgun boasts incredible range and an integrated dart holder, but hidden inside is a unique safety mouthpiece that allows the conical darts to slide into the blowpipe but not reverse out.

## Supplies

1 cereal box (17 ounces)
1 small paper clip
Electrical tape
Clear tape
1 drinking straw (optional)

## Tools

Safety glasses
Marker, pen, or pencil

Ruler
Large scissors
Hobby knife
Hot glue
Pliers

## Ammo

1+ square sticky note (3 inches by 3 inches)
1+ wooden skewer

# Step 1

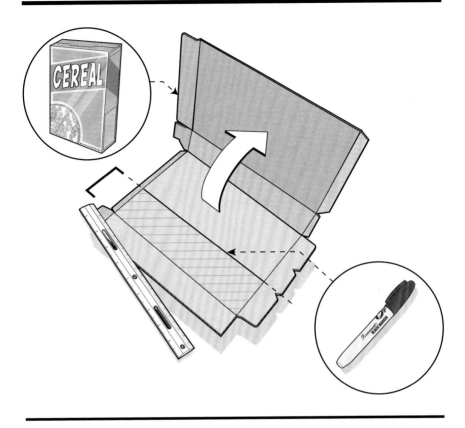

Start with a large, empty 17-ounce cereal box or cardboard box similar in size and width. Open up the box by carefully peeling apart the factory-glued tabs that hold the box together. The finished package should lie flat and completely open, with the graphics facing down.

Use a marker and ruler to mark off a 3-inch-by-box-height rectangle, as shown.

# Step 2

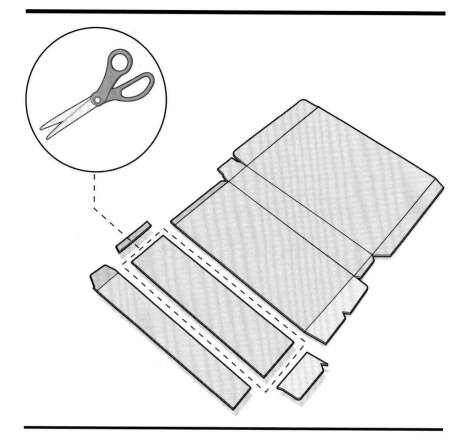

With scissors, remove the 3-inch-wide rectangle. Set aside the remaining pieces for optional blow dart mounts and additional blowguns or blowpipe extensions.

# Step 3

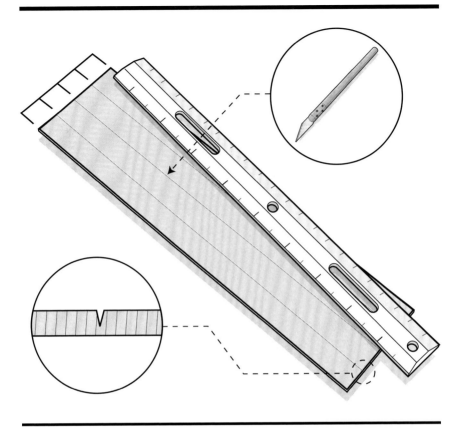

With ruler and marker, measure and mark four ¾-inch columns on the nongraphic side of the cardboard rectangle. Then, with a hobby knife, score the ¾-inch lines. *Do not* cut the lines through, but simply notch the lines with the blade, penetrating only half the width of the cardboard or less.

# Step 4

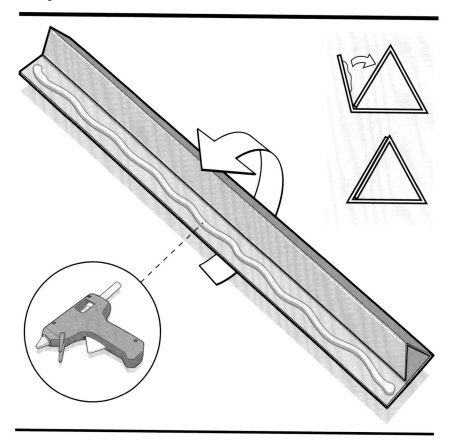

Using the cut score lines as a hinge, fold each panel 135 degrees downward to form a triangle, with the fourth column overlapping the first for added strength, as shown. Once folded, add hot glue between the first and fourth panels to stick them together. The final blow tube shape should be a triangle when viewed from the end.

# Step 5

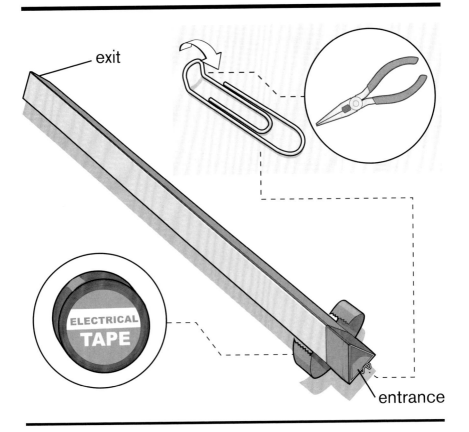

exit

ELECTRICAL TAPE

entrance

To create a safety mouthpiece, use pliers to bend one loop of a small paper clip upward 90 degrees, approximately ³⁄₁₆ inch from the end of the single-loop side. With electrical tape, attach the paper clip next to the blowpipe entrance, with the ³⁄₁₆-inch bend obstructing the entrance of the pipe. This unique design will allow the conical darts to slide into the blowpipe but not reverse out.

Put additional electrical tape around the entrance end of the triangle blowpipe to finish the protective mouthpiece. Without this added tape, over time the cardboard will lose its form due to dampness.

# Step 6

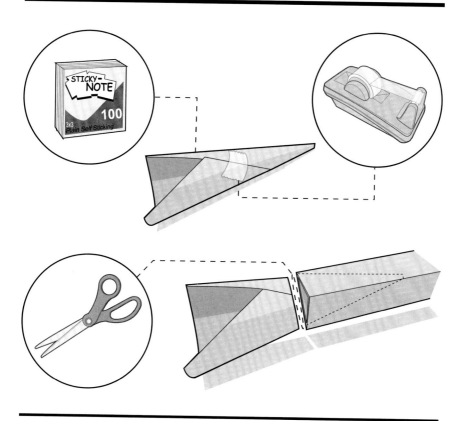

The Cereal Box Blowgun will fire a dart with a paper conical fletch. Roll one 3-inch-by-3-inch sticky note (or similar-sized piece of standard paper) into a funnel as shown. Use a small amount of clear tape to hold the paper in place. Then use additional sticky notes to make three more funnels if desired.

Next, insert one of the constructed funnels into the triangle-shaped blowgun (side without the paper clip), without damaging the tube or funnel. Once the funnel is gently wedged into the tube, use scissors to trim off the extra cone material hanging out of the blowpipe. The blow dart's maximum width is now the same as the blowgun, which will increase the contact surface area when you blow the dart through the tube. Repeat the trimming for the three remaining funnels.

# Step 7

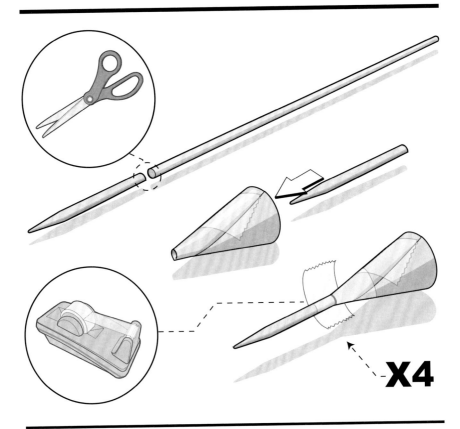

The point of the dart will be constructed from one wooden skewer. Cut the skewer with a pair of large scissors to be roughly 2 inches long.

Next, slide the pointed end of the skewer into the back of the funnel. Once it fully protrudes from the front, place tape around the paper cone to secure the skewer firmly in place. Repeat this step until all four recommended darts are completed.

To increase the blow dart's directional accuracy (optional), additional tape can be placed near the skewer point. You may need to experiment with the amount of tape.

# Step 8

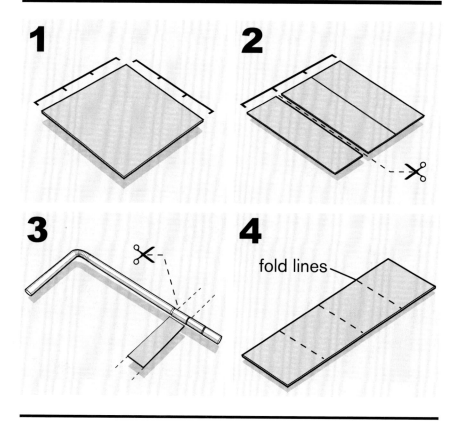

At this point, the Cereal Box Blowgun is fully functional. However, with a few added steps, optional dart holders can be constructed, making multiple-shot firing easy!

With scissors, cut out one 3-inch-by-2¼-inch cardboard rectangle (or similar size) from the remaining cereal box cardboard (1). Divide that rectangle into three equal strips, each ¾ inch wide and 3 inches long (2).

Use the width of the cardboard strips as a guide and cut three ¾-inch pieces from a standard drinking straw (3).

Prefold the strip into four sections, with three evenly spaced crease lines. If the cardboard strip's length is the recommended 3 inches, each crease line is at a ¾-inch interval (4).

# Step 9

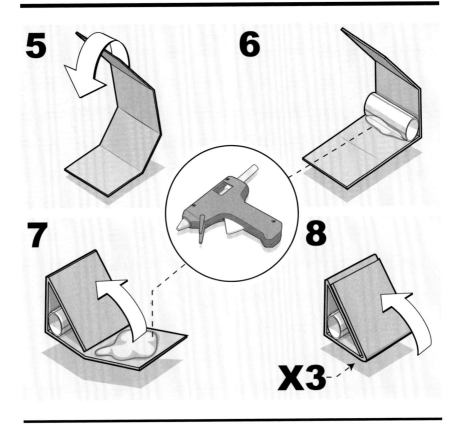

Fold the center crease at a 90-degree angle (5), then hot glue one straw cylinder in the inside corner of the cardboard strip (6).

Continue folding the strip to form a triangle, with the fourth section overlapping the first for added structure. Add hot glue on the inside of the fourth section (7), then fold and secure it to the first panel (8). Create three of these triangle assemblies, one for each side of the blowpipe.

# Step 10

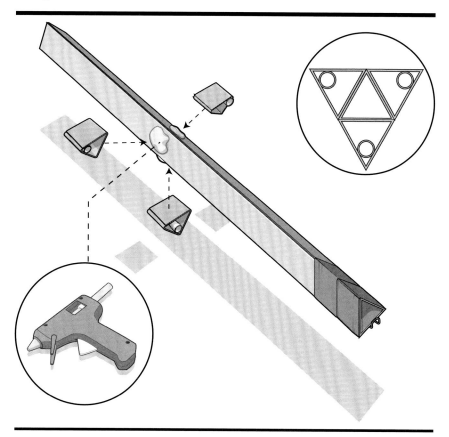

Hot glue each dart holder to one side of the triangle blowpipe, prefer-ably in the middle of the blowpipe. Prior to attaching the holders, make sure the straw is elevated away from the blowpipe frame to allow for blow dart clearance.

# Step 11

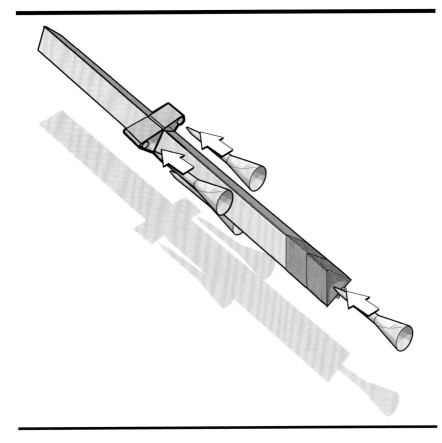

Fasten the skewer darts to the attached holders by inserting the dart tip until the conical fletch wedges into the straw, creating enough pressure to hold each dart in place.

To fire, load one of the darts into the entrance of the blowgun, skewer point first. Push it past the safety mouthpiece to secure it in the barrel. Now take a deep breath, lift the blowgun to your lips, aim, and exhale sharply to blow the dart out.

It is very important to remember that you're firing darts from your mouth. ***You should never inhale while the blowgun is in your mouth waiting to be launched.*** Always be responsible and fire the blowgun in a controlled manner. ***Wearing eye protection is a must, as is staying clear of spectators.*** Firepower is limited by your respiratory muscles, and accuracy is limited by the blowgun's length.

# DOUBLE PEN BLOWGUN

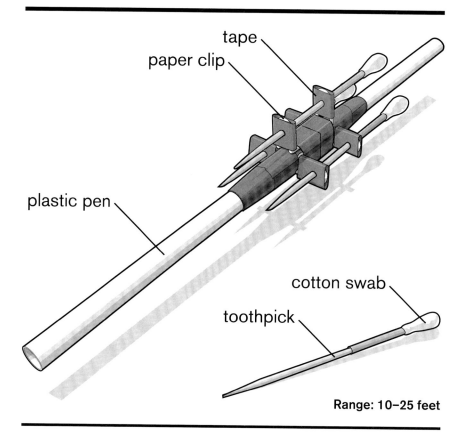

tape

paper clip

plastic pen

cotton swab

toothpick

**Range: 10–25 feet**

The easy-to-build Double Pen Blowgun, also known as the Double Pen Fukiya (Japanese blowgun), is a short-length competition blowgun built from two inexpensive ballpoint pens. In addition, the recommended darts are crafted from toothpicks and cotton swabs, supplies that are sold cheaply in bulk.

## Supplies

2 plastic ballpoint pens
Electrical tape
3 round wooden toothpicks
3 large paper clips

## Tools

Safety glasses

Pliers
Hobby knife (optional)
Scissors

## Ammo

1+ round wooden toothpicks
1+ plastic cotton swabs

# Step 1

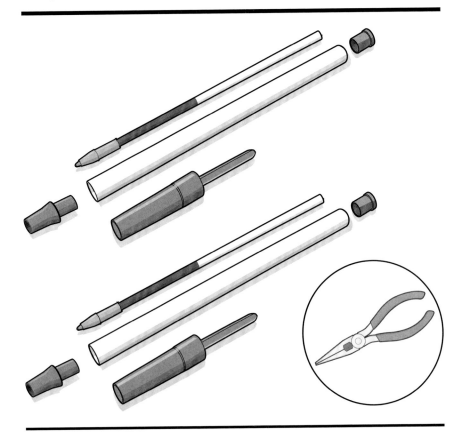

Disassemble two plastic ballpoint pens into their various parts. Depending on how the pens were manufactured, you may need a tool to assist you with dislodging the rear pen-housing cap. A pair of small pliers (for pulling it out) or hobby knife or large scissors (for cutting it off) work well.

# Step 2

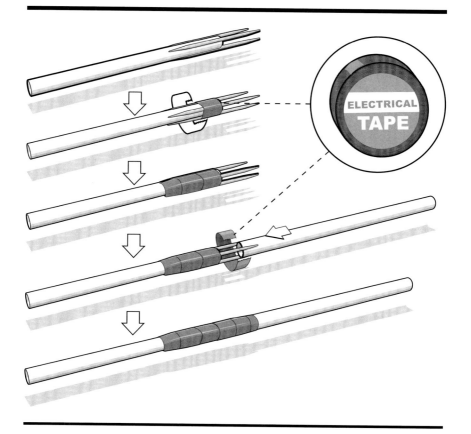

Using electrical tape, fasten three round wooden toothpicks, evenly spaced, onto the end of one pen housing. Each of the toothpicks should overhang halfway. Do not obstruct the pen tube with the added tape.

Slide the second pen housing between the attached toothpicks, aligning it with the first housing to make a seamless tube. Tightly wrap the pen housing in place with the tape.

# Step 3

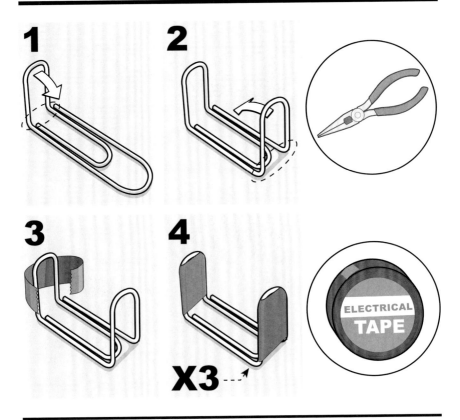

The optional dart holders can be constructed quickly and prove useful when target practicing. With pliers, bend both ends of a large paper clip upward at a 90-degree angle, approximately ½ inch from the tips (1 and 2).

Tightly tape around both paper clip bends as illustrated (3 and 4). This tape will hold the finished blow darts in place. Repeat this step until you have the recommended three modified paper clips.

# Step 4

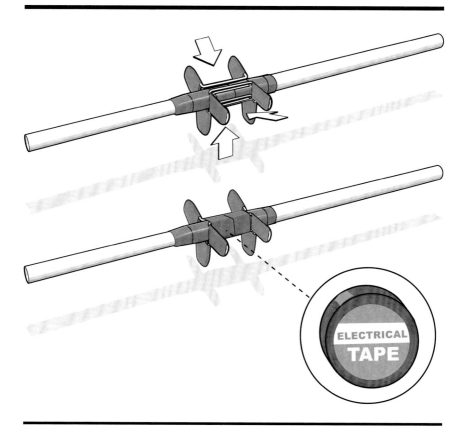

Attach the three modified paper clips, evenly spaced, to the center of the pen blowpipe by taping around the flat middle frame of the paper clips.

# Step 5

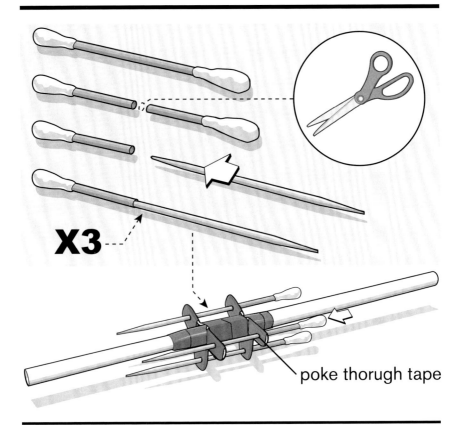

**X3**

poke thorugh tape

Each blow dart will be constructed from one plastic cotton swab and one toothpick. Cut the swab in half as shown and then slide a single round wooden toothpick into the swab's plastic cavity. Push the toothpick in until it's tight. You can glue the pieces together, but it's probably not necessary. Repeat to create three darts or more.

Finally, poke the wooden toothpick through both taped paper clip bends to hold the darts in place. You may want to make the hole nearest to the blowpipe entrance slightly larger for quick retrieval, and have the toothpick snug in the second hole only.

To fire, load one of the darts into the entrance of the blowgun, point first. Take a deep breath, lift the blowgun to your lips, aim, and exhale sharply to blow the dart out. ***Never inhale while the blowgun is in your mouth. Eye protection is a must, as is staying clear of spectators.*** Firepower is limited by your respiratory muscles, and accuracy by the gun's length.

# PLAYING CARD BLOWGUN

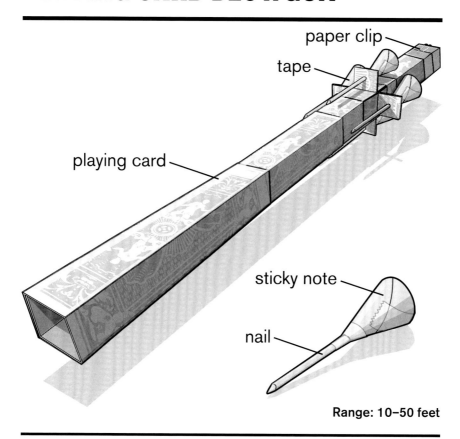

paper clip

tape

playing card

sticky note

nail

**Range: 10–50 feet**

Also known around the dojo as the Hanafuda (playing cards) Blowgun, this MiniWeapon has a very distinct square form and an added safety mouthpiece. Armed with nail darts, this blowgun is best suited for aluminium cans, Styrofoam, or wooden targets.

## Supplies

5 playing cards
Clear tape
1 small paper clip
Electrical tape

## Tools

Safety glasses
Scissors
Pen

Ruler
Hobby knife
Superglue or hot glue gun
Pliers

## Ammo

1+ trim nail (2 inches long)
1+ square sticky note (3 inches by 3 inches)

# Step 1

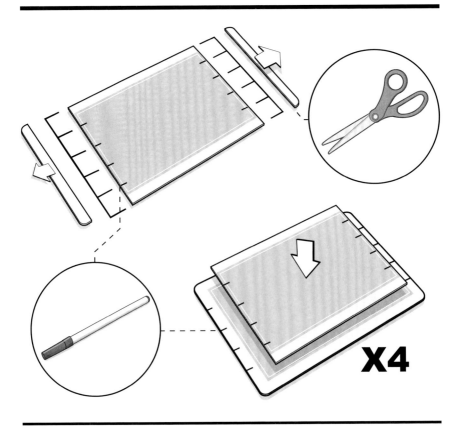

**X4**

To assist in construction of the Playing Card Blowgun, the first step is to create a folding template. With scissors, trim 1/8 inch off both width edges of one playing card. Then, divide both width edges of the card into five sections, approximately 7/16 inches apart, using a pen to mark these divisions. Both marked sides should be parallel to one another. This card has now been transformed into a folding template.

Place the card template on top of four additional cards and transfer the pen lines. When completed, all four cards share the same set of lines. Do not discard the template; it will be used again in step 5.

# Step 2

**X4**

With ruler and hobby knife, score all four marked cards (not the template). Do not cut the lines through, but simply notch the lines with the blade penetrating only half the width of the cardstock or less. The added score lines will act like a hinge, making folding the card easier.

# Step 3

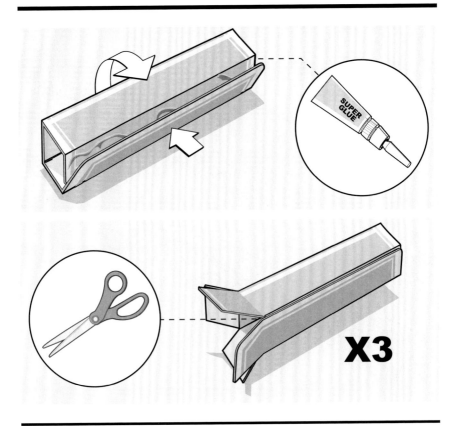

**X3**

Using the score lines as a hinge, fold each panel inward 90 degrees to form a box, with the fifth panel overlapping the first for added strength. Once folded, superglue or hot glue between the first and fifth panels and stick them together. Repeat this step for the remaining three cards, for a total of four finished boxes.

When the card assemblies are dry, modify three of them by cutting a ½-inch slit in all four corners along the vertical edge. On all three cards, fold those flaps open as shown in the bottom illustration.

# Step 4

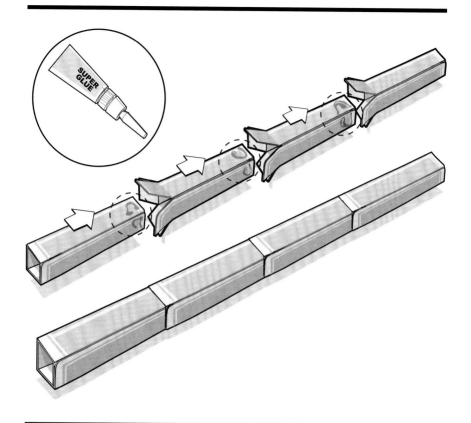

It's time to combine all four cards with superglue or hot glue into a working blowpipe. Start by adding glue to the ½-inch end of the uncut card box and sliding the ½-inch flaps from another card box over the assembly. When the glue is dry, add the next card box flaps onto the last card box with flaps, and so on until all four card boxes are assembled as shown.

Please note: The entrance of the Playing Card Blowgun will be at the end with the card box without flaps. If you don't use this end, the dart will bang against the interior card steps, causing misfires.

# Step 5

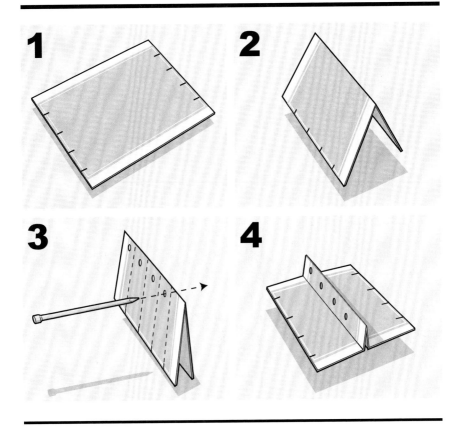

Locate the playing card template from step 1 and then fold the card in half to reduce its height (1 and 2).

The darts will be fashioned out of 2-inch-long molding and trim nails (small-headed nails). Use one of these nails as a hole punch—with the card in half, use the guidelines to evenly poke four holes through the folded cards, roughly ¼ inch from the crease line and centered between each pen mark indicated by the dotted lines (3).

Fold the card twice more ½ inch below the crease line with each fold going in a separate direction (4).

# Step 6

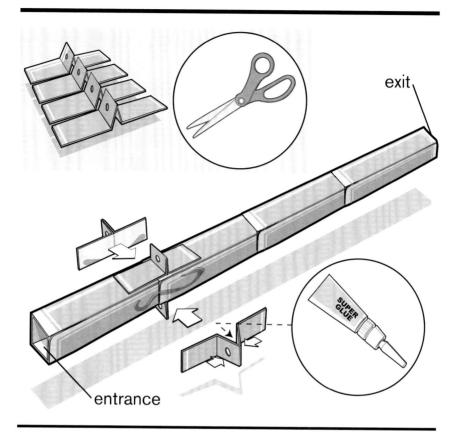

Use the marked lines (dotted lines from step 5) on the folded card to cut out each strip, with folds and the nail hole in the center of each (top).

Next, superglue or hot glue all four cards to each side of the square blowpipe 2½ inches from the entrance end (card box without the cut flaps).

# Step 7

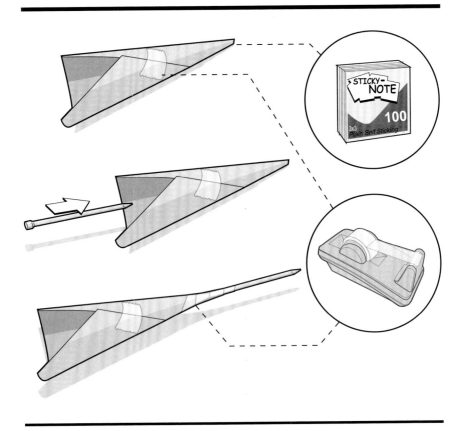

Now manufacture the nail blow darts. Roll one 3-inch-by-3-inch sticky note into a funnel as shown. Use a small amount of clear tape to hold the paper funnel in place.

Next, slide the pointed end of the nail into the back of the funnel. Once it fully protrudes from the front, tape around the cone to secure the nail firmly in place.

# Step 8

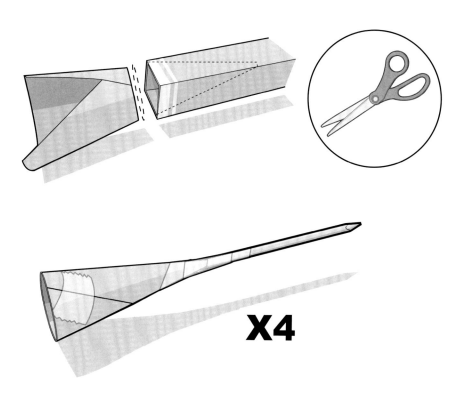

**X4**

Insert one of the dart assemblies, point first, into the playing card blow-pipe, without damaging the funnel. Once the funnel is gently wedged into the tube, use scissors to trim off the extra cone material hanging out of the tube. The blow dart's maximum width is now the same as the width of the blowpipe, which will increase the contact surface area when you blow the dart through the tube. Repeat steps 7 and 8 to create the remaining four darts.

# Step 9

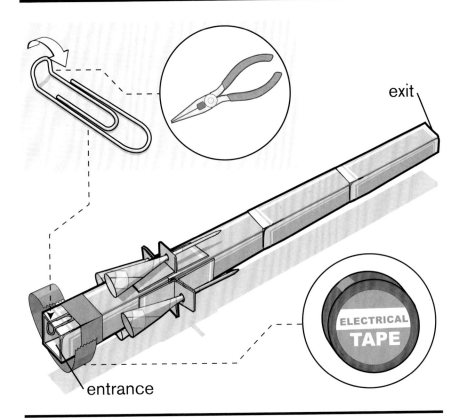

To create a safety mouthpiece, use pliers to bend one loop of a small paper clip 90 degrees, about ³/₁₆ inch from the end of the single-loop side. With electrical tape, attach the paper clip next to the pipe entrance, with the ³/₁₆-inch bend over the opening. This unique design will allow the conical darts to slide into the blowpipe but not back out.

Put additional electrical tape around the entrance end of the triangle blowpipe to finish the protective mouthpiece. Without this added tape, over time the cardboard will lose its form due to dampness.

To fire, load one of the darts into the entrance of the blowpipe, point first. Push it past the safety mouthpiece to secure it in the barrel. Take a deep breath, lift the blowgun to your lips, aim, and exhale sharply to blow the dart out. *You should never inhale while the blowgun is in your mouth. Eye protection is a must, as is staying clear of spectators.* Firepower is limited by your respiratory muscles, and accuracy by the gun's length.

# MINI TOOTHPICK BLOWGUN

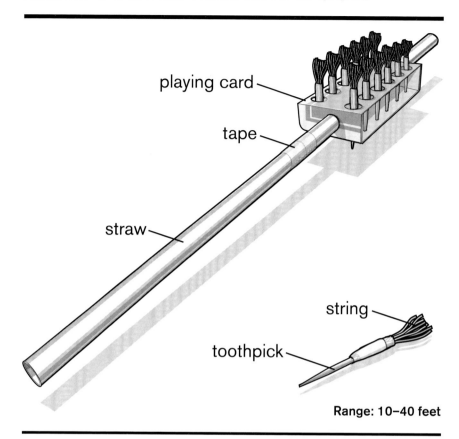

playing card

tape

straw

string

toothpick

**Range: 10–40 feet**

With a small bill of materials, the Mini Toothpick Blowgun is a great ninja project for a rainy day. Its mosquito-like darts are perfect for popping balloons! Because this blowpipe is constructed from drinking straws, a dedicated ninja has the option to use it as a snorkel, waiting submerged for hours until it is time to pounce.

## Supplies

2 drinking straws
Clear tape
1 playing card

## Tools

Safety glasses
Scissors

Ruler
Single-hole punch

## Ammo

6+ round wooden toothpicks
6+ pieces of fine string or
    thread (12 inches each)

# Step 1

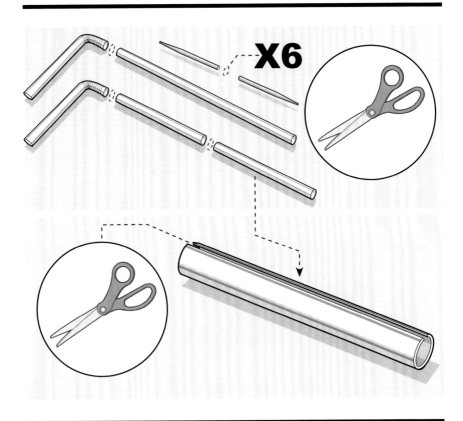

The first step in constructing the Mini Toothpick Blowgun is to prepare two drinking straws and toothpicks. If the straws have flexible heads, use scissors to cut them off, just below the bellowed (bendy) sections. Then cut one of the straws in half. In addition, cut all six toothpicks in half.

One of the drinking straw halves will need to be sliced up the middle. With the tip of the scissors inside the drinking straw, make one continuous cut along the length of the half straw section.

# Step 2

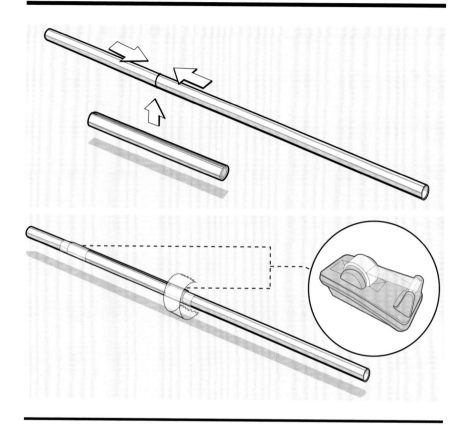

Align the long drinking straw and half drinking straw. Use the half cut segment to fasten both straw sections by snapping the cut slit over the two aligned sections (top).

Center the added half section over the joint and secure the assembly with clear tape.

# Step 3

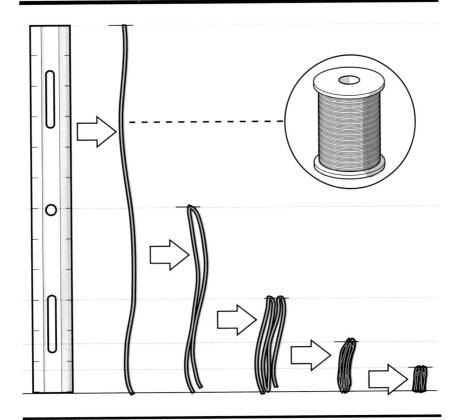

The dart tail will be constructed from fine string or thread. With scissors and ruler, cut a 12-inch length for each of the darts you plan on making; we suggest 6 to 12 darts. Then fold each section of string in half four times, reducing the string's length and at the same time significantly increasing its mass.

# Step 4

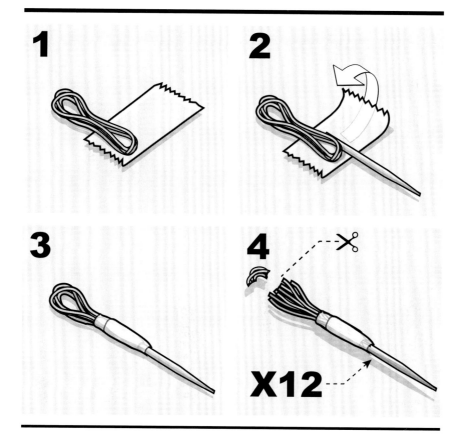

Place one section of folded string halfway on the end of a 1-inch piece of clear tape (1).

Next, add a cut toothpick to the tape, with the blunt end of the toothpick slightly overlapping the tape string (2).

Tightly roll the string and toothpick together, adding more tape if needed (3). With scissors, cut the rear string loops so the string is unattached at the rear (4). Repeat this step to create the recommended 6 to 12 darts.

# Step 5

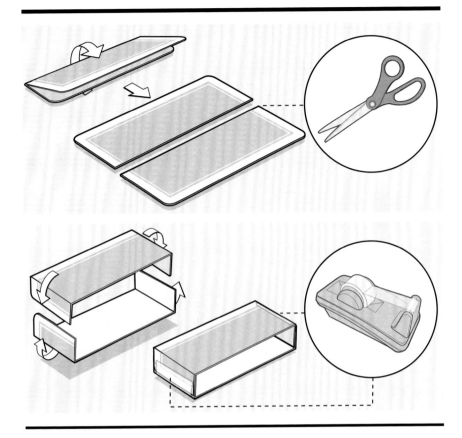

The optional dart holder makes multishot target practice easy. Start by folding one playing card in half vertically. Use the center crease line as a cutting guide and cut the card in half.

Next, ½ inch from both ends of the card segments, fold the ends upward 90 degrees. Then flip one card and slide them together to form a 3D rectangle. Put tape at the ends to hold the rectangle together.

# Step 6

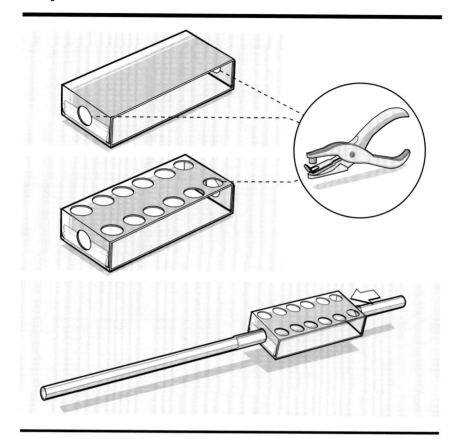

Using a single-hole punch, add one hole to each rectangle side, directly in the middle (top image).

Use the hole punch to add a set of holes on the top surface of the dart holder. We recommend two sets of six for a total of 12 holes. Each hole will represent one dart holder.

Next, slide the dart holder onto the straw blowpipe, with the assembly ending up over the doubled layered straw end for a snug fit. This end will be the blowpipe entrance.

# Step 7

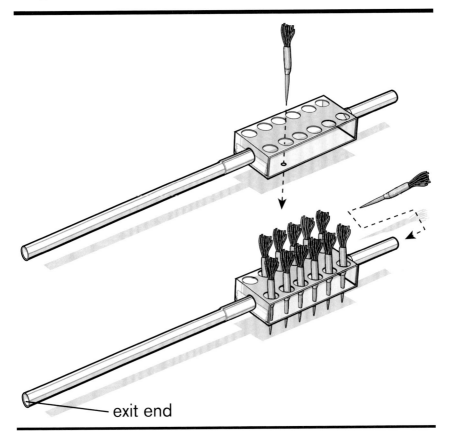

exit end

To secure the darts in place, insert one dart per hole punch opening, piercing the bottom card with the toothpick point to lock the dart in place. Continue to fill each dart holder until the ammo clip is locked and loaded for target practice.

To fire, load one of the darts into the entrance of the blowpipe, point first. Take a deep breath, lift the blowgun to your lips, aim, and exhale sharply to blow the dart out.

It is very important to remember that you're firing darts from your mouth. ***You should never inhale while the blowgun is in your mouth waiting to be launched.*** Always be responsible and fire the blowgun in a controlled manner. ***Wearing eye protection is a must, as is staying clear of spectators.*** Firepower is limited by your respiratory muscles, and accuracy is limited by the blowgun's length.

**Pointed Dart Blowguns**

## 5

# SIEGE WEAPONS

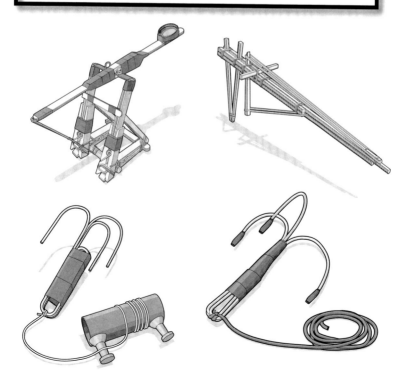

# JAPANESE ROCK THROWER

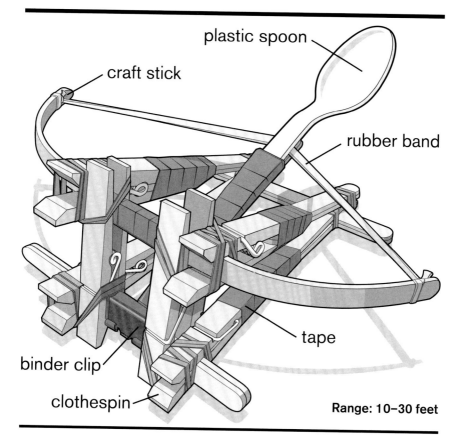

plastic spoon

craft stick

rubber band

tape

binder clip

clothespin

**Range: 10–30 feet**

Requiring no glue and hurling projectiles with unforgiving accuracy, this wartime miniature represents one of the most diverse siege weapons on the ancient battlefield, which could be modified to launch wall-crushing projectiles or shoot large antipersonnel arrows.

## Supplies

16 craft sticks
13 rubber bands
Electrical tape
6 wooden clothespins
1 medium binder clip (32 mm or 25 mm)
1 plastic spoon
1 wide rubber band

## Tools

Safety glasses
Hobby knife
2 2-peg-by-4-peg building blocks
Bowl of warm water
Large scissors

## Ammo

1+ soft candies

# Step 1

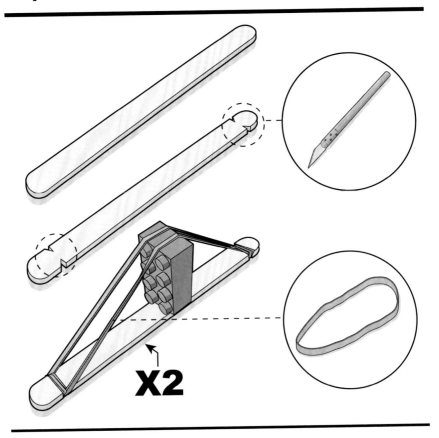

**X2**

With a hobby knife, carefully cut two small notches in both ends of a craft stick, approximately ½ inch from the tips. Then loop one rubber band around both groups of notches. Once the band is in place, slide a 2-peg-by-4-peg building block (or something equivalent in size) upright, under the rubber band. Repeat this step with another craft stick for a total of two finished assemblies.

# Step 2

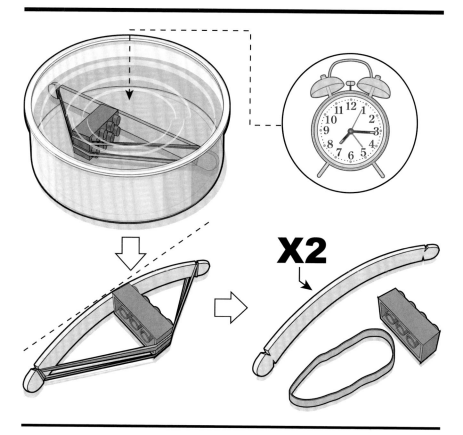

Bending wood with water is a fascinating process used by all types of woodworkers. You will be using this technique to give the craft sticks additional pliability. Fully submerge both craft stick assemblies in a full bowl of warm water. Soak the assemblies for 60 to 90 minutes before removing and bending them.

With the rubber band and block still attached, slowly increase each stick's arc with your fingers. Let the wet craft sticks dry for 30 minutes or more to allow the arc to set. Once the sticks are dry, remove the rubber bands and blocks.

# Step 3

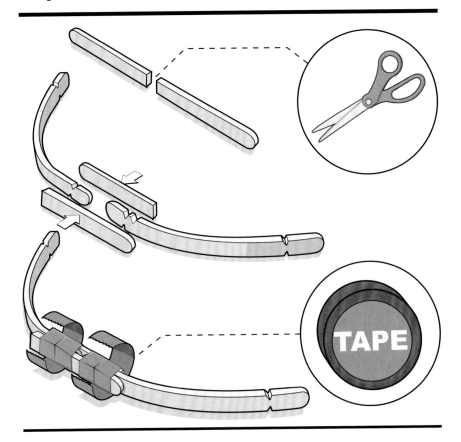

With large scissors, cut one craft stick in half as shown. Use both halves to sandwich both curved craft sticks from the previous step. When the ends of the sandwiched sticks are touching, use electrical tape to bind the entire assembly.

# Step 4

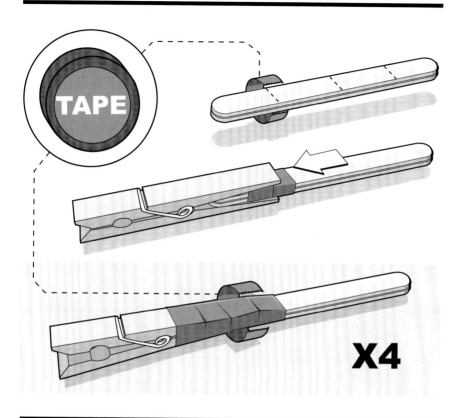

The next step is constructing four identical braces for the Japanese Rock Thrower. To begin, stack two craft sticks as shown. Then use tape at one end to secure the sticks (top image).

Slide the stick assembly between the rear prongs of one clothespin, inserting the taped end roughly 1 inch deep. Once in position, tightly tape the assembly together. Because these are supports, use plenty of tape to reduce movement.

Repeat this step until you have four identical assemblies.

# Step 5

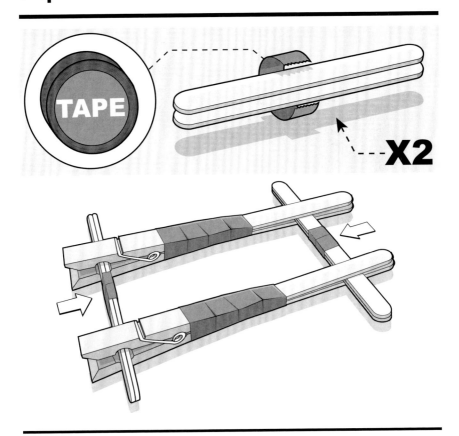

Begin building the lower frame by creating two more craft stick assemblies. Start by stacking two craft sticks as shown (top image), placing tape at the center to secure the sticks. Repeat this step for one additional craft stick group.

Clip one craft stick group into two wooden clothespin frames, with the stick protruding 1 inch from each end. This grouping should be positioned vertically (bottom left).

Sandwich the second craft stick group between the rear attached craft sticks fixed to the clothespin. This grouping should lie horizontally, with the ends sticking out 1 inch, just like the front (bottom right).

# Step 6

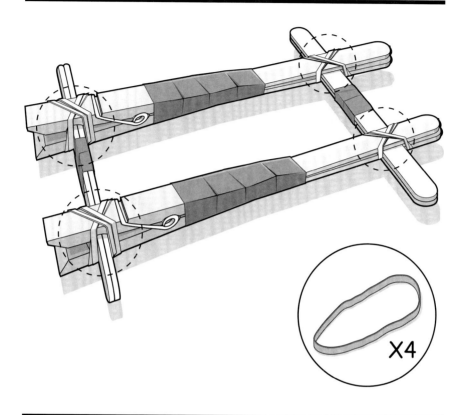

Use four rubber bands to fasten the four connection points on the frame.

# Step 7

do not tape

TAPE

With the lower frame complete, begin construction of the catapult arm. Locate one medium (32 mm or 25mm) or similar-size binder clip that can clamp onto the width of two craft sticks—test fit it now. Tape one craft stick onto the inside of the metal handle attached to the binder clip.

To add length and a projectile basket, tape one plastic spoon, facing up, on the top of the craft stick as illustrated. The handle should overlap the attached craft stick by 2 to 3 inches. *Do not* tape the tip of the craft stick (as indicated); this detail will be used to hold the elastic bow string in step 12.

# Step 8

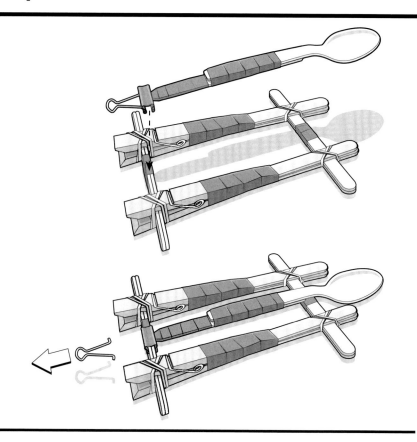

Using the binder clip, attach the swing arm assembly to the lower craft stick brace, positioned in the middle. Once in place, remove the front metal handle of the binder clip.

# Step 9

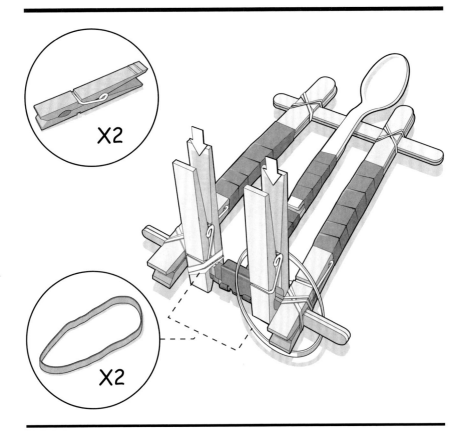

X2

X2

Attach two clothespins to the lower frame, clamped between the binder clip and both brace assemblies. Once they are aligned and at 90-degree angles, use two rubber bands to fasten the clothespins to the frame.

# Step 10

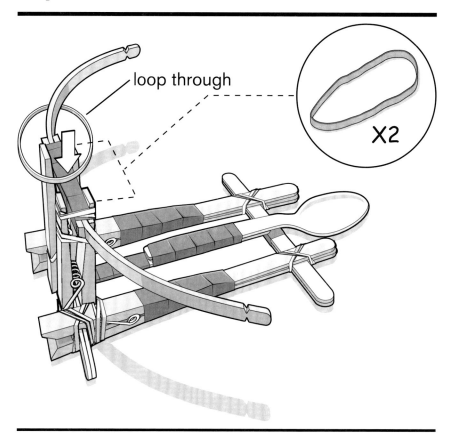

loop through

X2

Slide the bow assembly between the attached upright clothespin prongs. The bow should be centered between the prongs, with the bow curve arced in the direction of the swing arm.

When the assembly is in place, slide a rubber band around each end of the bow and secure it by wrapping the band around the prongs.

# Step 11

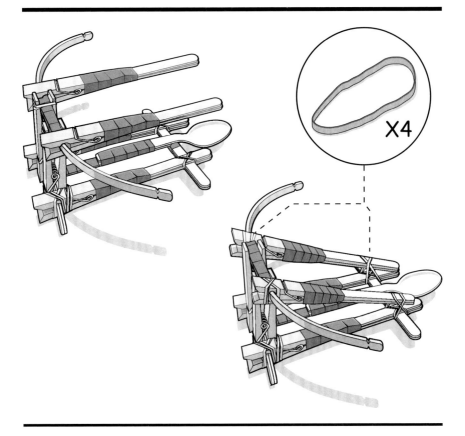

X4

Clip the two remaining braces onto the bow, directly above the lower frame braces.

Rotate the upper braces down so that the craft stick tips touch the lower frame. Use two rubber bands to fasten the lowered ends in place. Then use two additional rubber bands around the upper clothespins, fastening the clothespins to the attached bow.

# Step 12

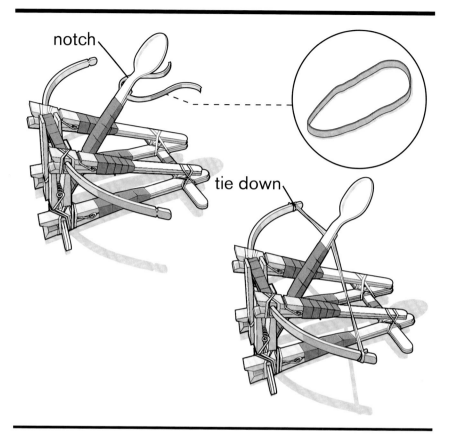

notch

tie down

It's time to add the elastic firepower to this rock thrower. With scissors, cut open a wide rubber band. Wedge the center of the rubber band between the craft stick and spoon neck (top image).

Next, straighten the rubber band and then tie each end of the rubber band onto opposite ends of the bow limbs, using the notches to help hold the rubber band knot. The rock thrower is now complete.

# Step 13

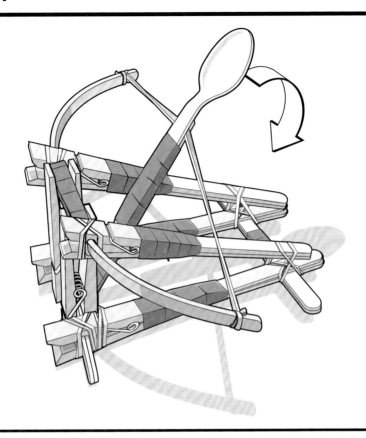

Now you're ready to fire. Pull down the catapult basket, load it with ammo, and release! This catapult is perfect for launching all types of projectiles, including coins, spitballs, erasers, and soft candies. ***Remember to use eye protection! Never aim this catapult at a living target, and use only safe ammunition.***

# HWACHA ROCKET CART

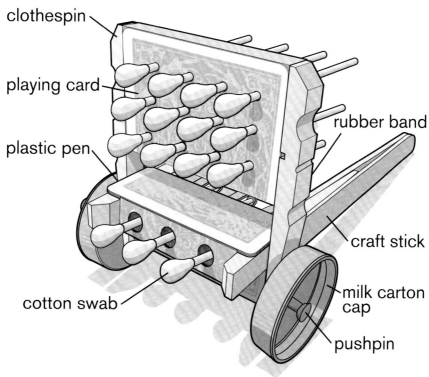

clothespin

playing card

plastic pen

rubber band

craft stick

milk carton cap

cotton swab

pushpin

**Range: 10–25 feet**

The original *hwacha* rocket cart was a technological breakthrough capable of firing up to 200 "fire arrow" rockets. The MiniWeapon version launches multiple cotton swab rockets, and its wooden frame can easily be repositioned to hold back dense ranks of imperial soldiers!

## Supplies

5 playing cards
2 wooden clothespins
8 craft sticks
1 plastic ballpoint pen
1 wide rubber band
2 pushpins
2 plastic milk carton caps

## Tools

Safety glasses

Large scissors
Glue stick
Hot glue gun
Pen
Ruler
Single-hole punch
Toothpick (or similar pointed tool)

## Ammo

15+ plastic cotton swabs

# Step 1

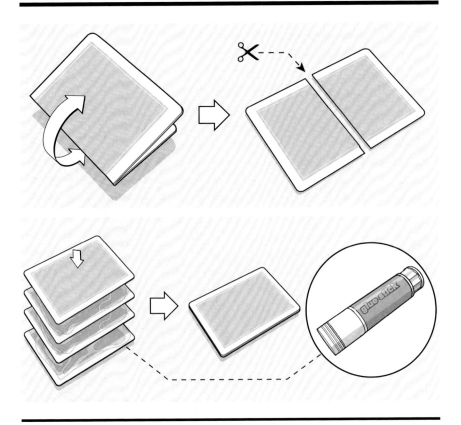

The Hwacha Rocket Cart panels will be constructed from multiple playing cards.

Fold one playing card in half and then open the card up and use the center crease line as a cutting guide. With scissors, cut the card in half and then set both parts aside until step 9.

Use a glue stick (recommended) or hot glue gun to attach four playing cards together in a stack. Let the glued cards dry completely before proceeding to the next step.

# Step 2

The glued card stack will be transformed into the cart panels of the Hwacha Rocket Cart. Using a pen and ruler, draw a line indicating the center of the card stack (1). Then, on one half of the card stack, draw a second line dividing the half into another half (2). On the opposite side, divide the card stack half into three equal parts, indicating each part with a pen line as shown (3).

# Step 3

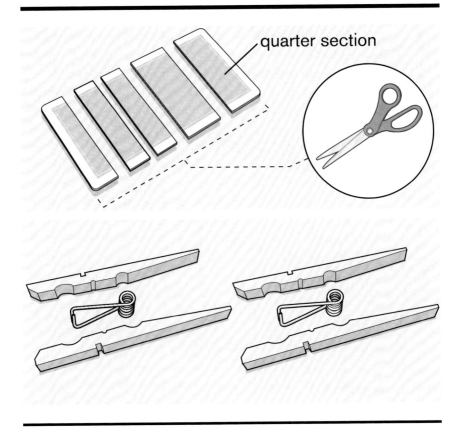

quarter section

Using the pen lines as guides, cut out each section of the card stack. The two quarter sections will be the top and bottom of the Hwacha Rocket Cart; the thirds will be used for the front and axle detail.

Now, disassemble two wooden clothespins. The springs will not be needed for this project–recycle them.

# Step 4

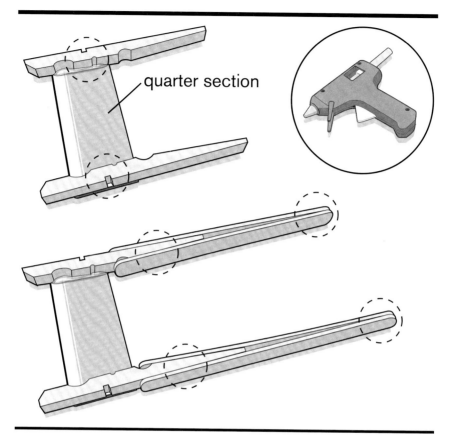

quarter section

To start the frame construction, hot glue two separated clothespin prongs along the width edge of one quartered section of cards. Position both wooden prongs ¼ inch past the cards, with the flat side of the prongs to the outside as illustrated.

Next, hot glue two craft sticks to the back side of each attached prong, against the attached card. First secure the craft sticks on the prongs and, once dry, use additional glue to fix the ends of the sticks together.

# Step 5

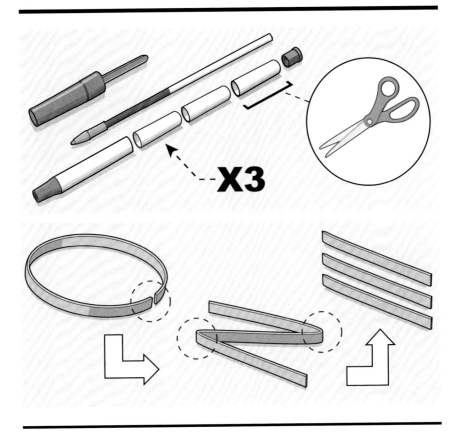

Disassemble one plastic ballpoint pen into its various parts. Dislodge the rear pen-housing cap and remove the ink cartridge. You may need a tool (pliers) to dislodge the rear pen-housing cap. Use large scissors to cut three 1-inch segments out of the pen housing. Set these pieces aside for the next step.

With scissors, cut open a wide rubber band. Then fold that rubber band twice, to divide the length by one-third. Cut the rubber band at each crease so you are left with three identical-length rubber band segments.

# Step 6

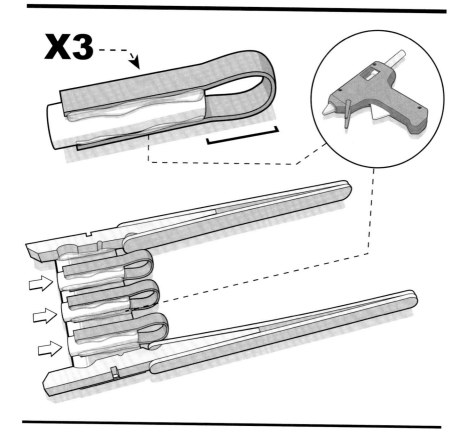

The Hwacha Rocket Cart can fire three cotton swab rockets in succession. To make this happen, three identical firing barrels will be created from the cut rubber band and pen segments from step 5. Hot glue one rubber band segment to the top and bottom of the 1-inch pen segment, with the back rubber band loop hanging ¾ inch over the back and looped back onto the pen. This ¾ inch is important for easy pullback. Complete this step for two additional barrel assemblies, as shown.

Hot glue all three barrel assemblies, evenly spaced, between the fixed clothespin prongs, with the front pen segment flush to the card edge. Once attached, there should be roughly ¼ inch space between each barrel assembly and fixed side prongs.

# Step 7

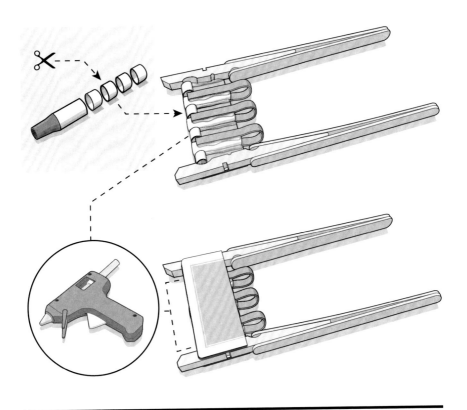

Use the remaining pen housing from step 5 to cut five small spac-ers to go between the fixed barrel assemblies. Each pen housing spacer should be around ¼ inch long and hot glued flush to the barrel entrance. These fixed spacers will give the Hwacha Rocket Cart addi-tional strength when the barrels are being fired.

Cap the assembly by hot gluing the remaining half-card segment on top, with the card edges aligning with the lower card assembly.

# Step 8

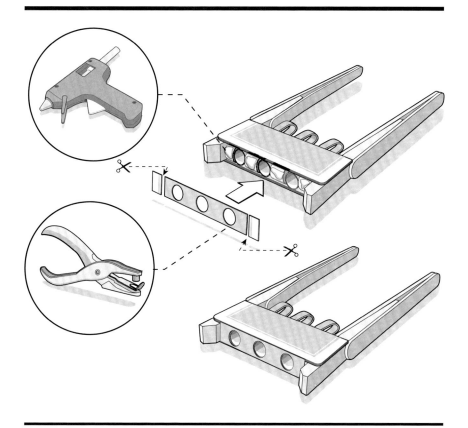

Modify one of the one-third-card sections to fit over the front of the cart (barrel exit). Use scissors to trim the width and a single-hole punch tool to make the three barrel exit holes. Once cut and ready, use hot glue to fix the one-third-card section to the front of the cart frame.

# Step 9

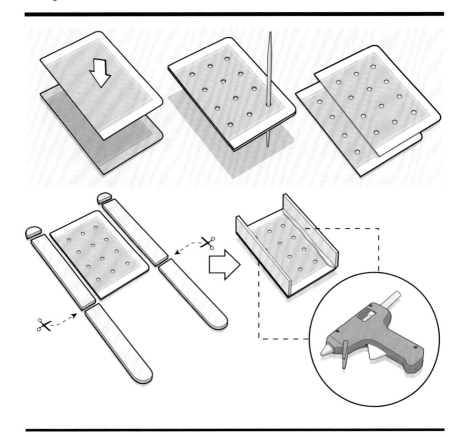

The card you cut in half in step 1 will be modified to hold additional soft tip rockets (cotton swabs). Stack the two halves on top of one another. Then use a tool such as a pointed toothpick to poke the recommended 12-hole pattern (4 holes across and 3 down) as shown, without going too close to the edge. When finished, you will have two card halves with identical hole patterns.

Next, cut two craft sticks to the length of the original card width (bottom left illustration). Hot glue both craft stick sections vertically 90 degrees along the original width edge of the card segment.

# Step 10

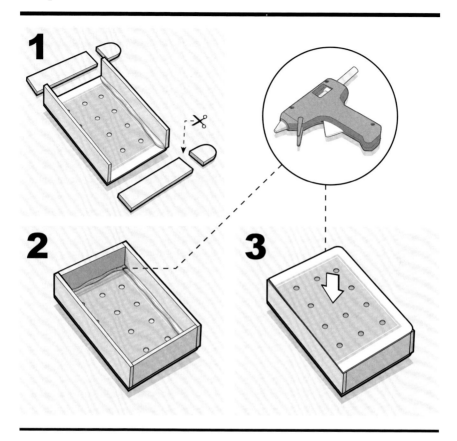

Cut the remaining craft sticks from step 9 to the new width of the half card (1). Slide the two cut craft stick sections vertically and hot glue them into place (2). Align the holes of the cards and then hot glue the remaining half-card section on top of the box assembly (3).

# Step 11

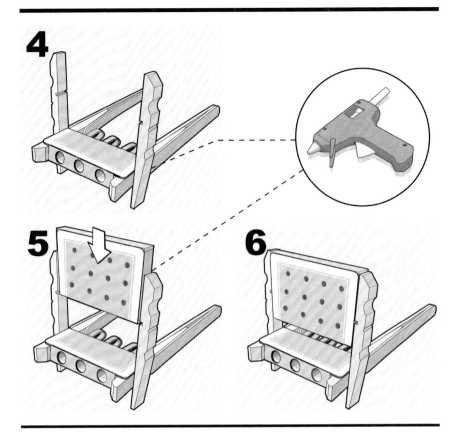

Hot glue the remaining clothespin prongs to the side of the frame assembly, with the flat surface inward. Position each prong 90 degrees with ¾ inch of the prong overhanging the lower frame and ½ inch back from the front edge of the card fixed to the frame (4).

The box assembly spacing should be identical to the space between the fixed prongs because both are based on the original playing card width. Hot glue both sides of the box assembly between the two prongs, leaving a ½-inch space between the frame and box (5 and 6).

# Step 12

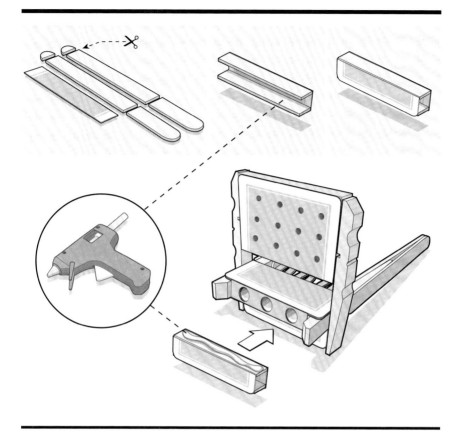

With a few quick steps you can add some authenticity to your Hwacha Rocket Cart. The last two card segments from step 3 can be used to add optional axle detail. To start, cut two craft sticks the same length of the original width of the playing card. Hot glue both craft stick segments to opposite edges of the playing card. Then cap off the box by gluing the last card segment to the top of the craft sticks. The axle detail is finished.

Hot glue the finished axle detail under the frame, positioned between the two upright prongs.

# Step 13

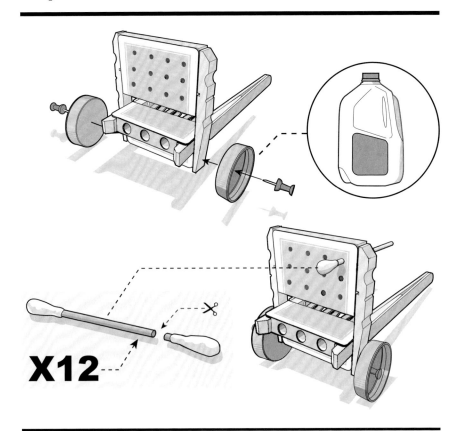

**X12**

To add wheels to the Hwacha Rocket Cart frame, center two pushpins through two plastic milk jug caps. Then attach the wheels by pushing the pushpin point into the upright clothespin prong.

With scissors, cut the tips off 12 cotton swabs to make plenty of cotton swab rockets. Load each cotton swab into the 12-slotted rocket holder.

To fire your Hwacha Rocket Cart, remove one of the cotton swab rockets and load it into the pen housing opening, pointed swab side out. Find the cut end of the cotton swab in the rubber band, and then pull back the rubber band and release. ***Remember to use eye protection!*** Cotton swab rockets (arrows) have sharp points and can travel at a high velocity. ***MiniWeapon projects are not meant for living targets.*** Always stay clear of spectators and operate the bow in a controlled manner. Homemade weaponry can malfunction.

# CROUCHING TIGER CATAPULT

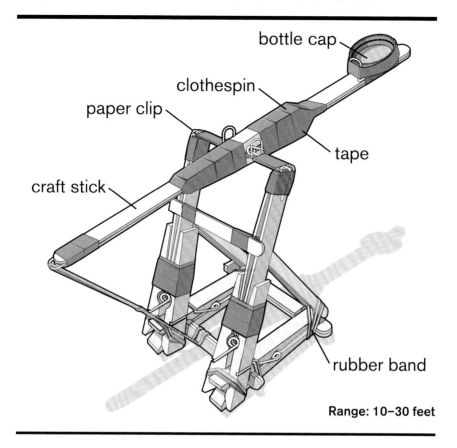

bottle cap

clothespin

paper clip

tape

craft stick

rubber band

**Range: 10–30 feet**

Inspired by the battlefields of the Tang dynasty (AD 618–907), this medium-sized traction catapult is perfect for launching all types of projectiles, including coins, spitballs, erasers, and marshmallows. With a tall stance and wicked swing arm, this catapult's silhouette will strike fear in the opposition.

## Supplies

13 craft sticks
Electrical tape
5 wooden clothespins
10 rubber bands
7 large paper clips
1 plastic bottle cap

## Tools

Safety glasses
Large scissors
Pliers
Ruler (optional)

## Ammo

1+ soft candies

# Step 1

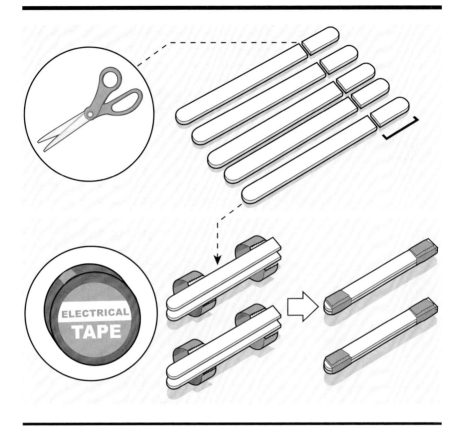

Part of the lower frame of the Crouching Tiger Catapult will be made up of two craft stick supports. With large scissors, cut five craft sticks to a length of 3½ inches. Four of the cut sticks will be used in this step; the fifth cut stick will be used in step 6 as a cross brace.

Divide the four cut craft sticks into two stacks. Then fasten both ends of each stack with electrical tape.

# Step 2

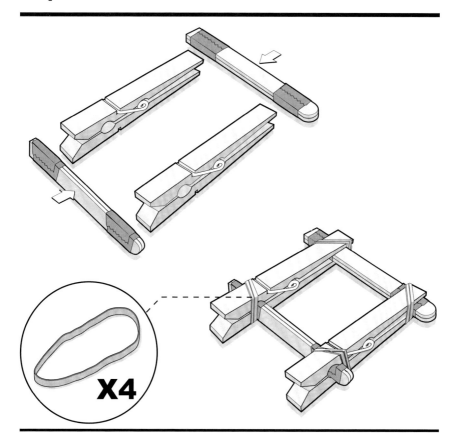

**X4**

Clamp one craft stick group into the round detail of two wooden clothespins, with the clothespins positioned on the tape and the craft sticks vertical. Slide the second craft stick group between the rear prongs, inserted ½ inch in. Use one rubber band at each intersection to secure the lower frame assembly. Both craft stick groups should stick out ¼ inch.

# Step 3

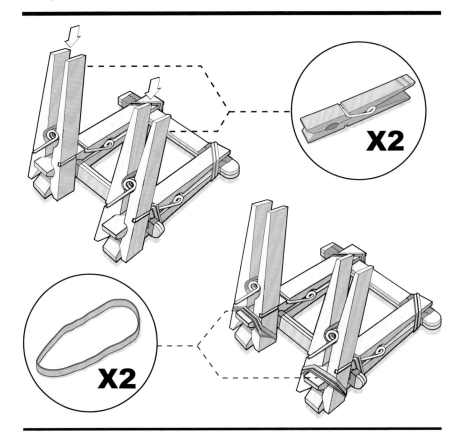

Clamp two additional clothespins to the prong tips of the attached clothespins on the lower frame; clamp each of them at an approximately 75-degree angle inward. This angle will adjust slightly as the build continues. Use two rubber bands to secure both clothespins to the lower frame.

# Step 4

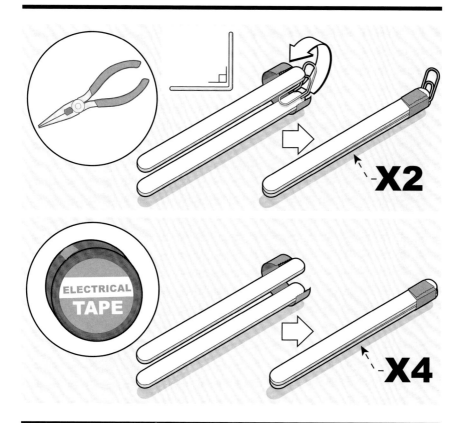

The Crouching Tiger Catapult has a tall frame that will require you to make additional supports and height extensions from craft sticks. With pliers, bend two large paper clips, each at a 90-degree angle at the center as shown. Sandwich one end of the bent paper clip between the end of two craft sticks. Secure the assembly by taping only the end with the paper clip. Repeat this step to complete two finished craft sticks with attached paper clips.

Divide eight more craft sticks into four groups. Tape each bundle of two at one end. Repeat this step until you have four completed bundles.

# Step 5

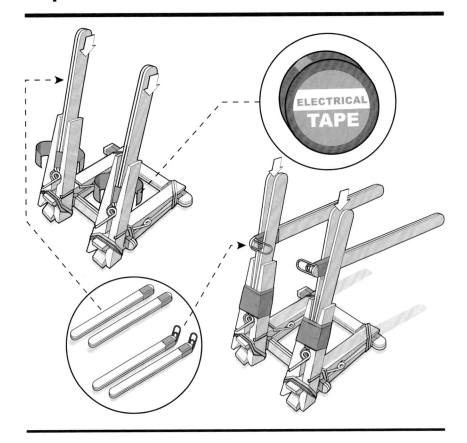

Taped end first, slide two craft stick assemblies without the attached paper clips between the two upright clothespin prongs. Once aligned with the attached clothespins, secure both craft stick assemblies with tape.

Next, wedge each of the craft stick assemblies with attached paper clips between the upright craft sticks on the frame. Slide them both down until they are near the end of the prongs.

# Step 6

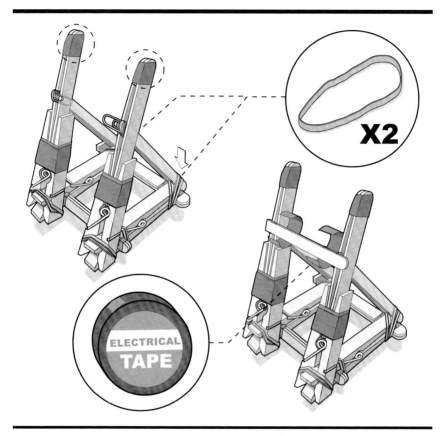

X2

ELECTRICAL
TAPE

Add tape around both ends of the upper craft stick assemblies. Then, angle the back ends of the two wedged braces to the rear of the lower frame. Touching the clothespin prongs, at around 45 degrees, use two rubber bands to hold the braces in place.

With tape, attach the fifth cut craft stick from step 1 to the two bent clothespins attached to the angled braces. This cross brace will add much-needed support and a stop for the catapult's swing arm.

# Step 7

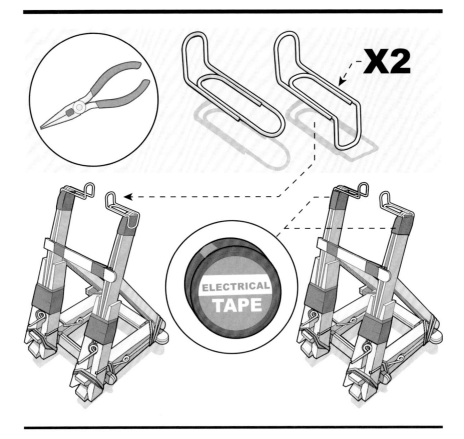

The Crouching Tiger Catapult's swing arm hinge mount will be constructed from two large paper clips.

With pliers, make one 90-degree bend at both ends of a large paper clip, each bend located ¼ inch from the end loop and opposite one another. Repeat this step to create one additional paper clip with the same bends.

Position and tape each bent paper clip facing each other on top of the catapult frame, using the craft stick uprights as mounting points. The angle detail on each paper clip should be to the outside of the craft sticks.

# Step 8

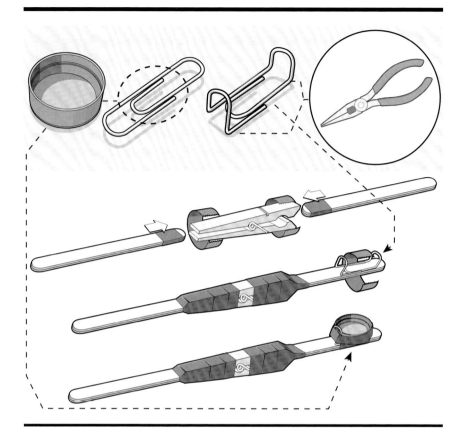

It's time to construct the swing arm and swing arm basket. With pliers, make two 90-degree bends in the large paper clip, using the bottle cap as a guide; the distance between the bends should be the same as the cap's diameter. The bent paper clip will act as a cradle for the bottle cap and will add much-needed support to the swing arm basket.

Slide the taped ends of both craft stick assemblies from step 4 into the front and back of the clothespin. Once aligned, tape the craft sticks in place.

Next, tape the bent paper clip ¼ inch from the tip of the wooden swing arm. Once in place, sandwich the plastic cap between the attached paper clip bends and tape the cap into place.

# Step 9

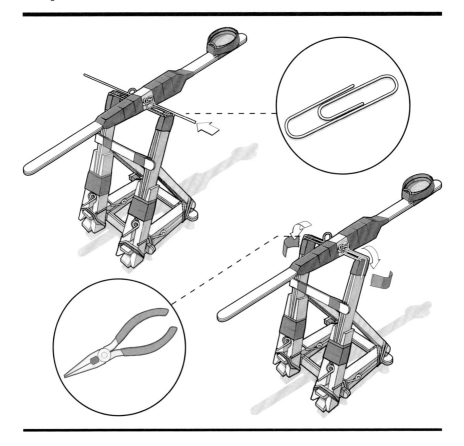

With pliers, straighten one large paper clip. Then slide the straightened rod through the mounted paper clips and spring located on the swing arm assembly.

Bend down both ends of the straightened paper clip at 90-degree angles, resting on the wooden craft sticks. Then use tape to fasten the bent paper clip rod to the catapult uprights. The swing arm should swing back and forth freely.

# Step 10

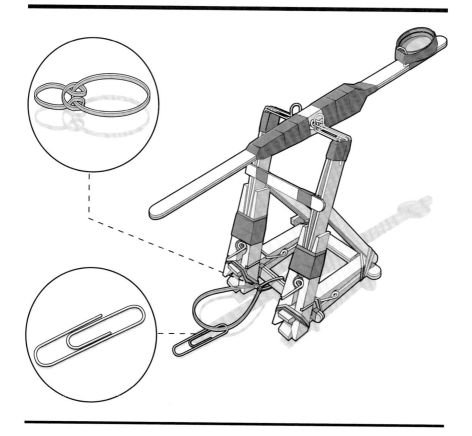

Loop two rubber bands together to construct one large double rubber band as shown. Then loop the double rubber band around the lower frame, centered between the upright clothespins, then back through its loop to secure the band to the craft stick brace.

Attach a large paper clip to the outer rubber band loop.

# Step 11

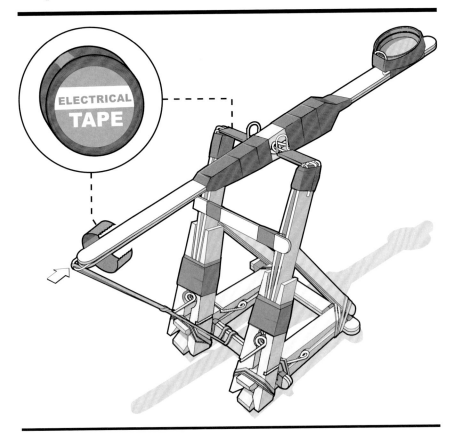

ELECTRICAL TAPE

Slide the rubber band–attached paper clip between the two lower craft sticks mounted on the swing arm. Once the paper clip is wedged in place, secure it by tightly wrapping tape around the assembly. Add additional tape around the two 90-degree bent paper clips for added structure.

To fire, hold down the bottom of the frame with one hand, pull back the loaded swing arm, and release when ready! **Remember to use eye protection! Never aim this catapult at a living target, and use only safe ammunition.** Soft mints and mini marshmallows work nicely.

# CHOPSTICK RUBBER BAND GUN

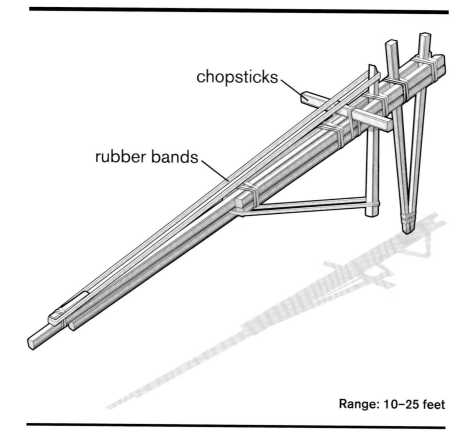

chopsticks

rubber bands

**Range: 10–25 feet**

The Chopstick Rubber Band Gun reminds us that nontraditional Mini-Weapons are a ninja's forte—everyday objects can be transformed into menacing tools! Constructed from wooden chopsticks and rubber bands, this gun shoots with impressive accuracy. With a nonlethal bill of materials, it's the perfect dojo trainer.

## Supplies

4 unseparated sets of disposable chopsticks
9 rubber bands

## Tools

Safety glasses
Large scissors or wire cutters
Ruler (optional)

## Ammo

1+ rubber band

# Step 1

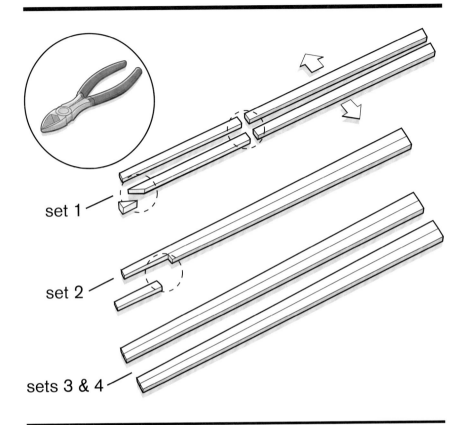

set 1

set 2

sets 3 & 4

Start with preparing four sets of disposable chopsticks, still attached at the base:

*Chopsticks Set 1:* Separate the chopsticks. With large scissors or wire cutters, cut 3 inches off both chopsticks. Cut a 45-degree tip on one of the 3-inch sticks, as shown. Save the remaining 5-inch segments as well.

*Chopsticks Set 2:* Keep this set attached. Cut 1 inch off the end of just one attached chopstick. Save the removed 1-inch stick.

*Chopsticks Sets 3 & 4:* Both sets will remain attached.

# Step 2

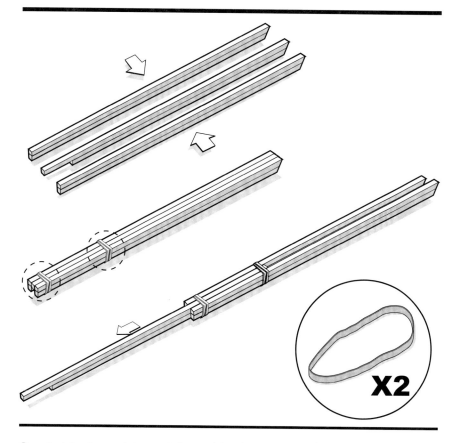

Sandwich chopstick set 2 (set with missing 1 inch) between chopstick set 3 and set 4. All three profiles should be the same.

To secure the assembly, use two rubber bands. The first rubber band should be located 1 inch from the tip; the second rubber band should be wrapped 3 inches from the same tip.

Once assembled, slide the center chopstick set (set 2) forward, until it extends to the second rubber band.

# Step 3

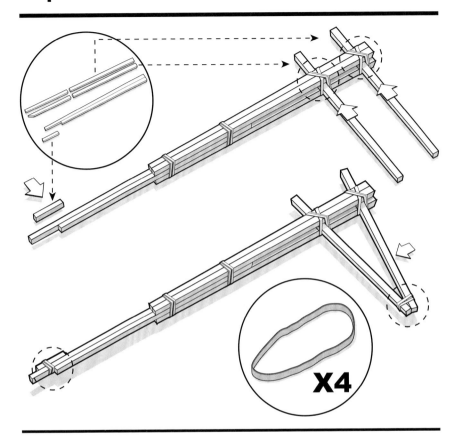

Three parts will be added to the chopstick frame in this step.

At the tip of the barrel, place the 1-inch stick opposite the notched-out chopstick. Use one rubber band to secure the 1-inch stick ½ inch from the tip.

Next, slide both 5-inch sticks from set 1 between set 3 and set 4. Use two rubber bands to fasten them to the frame 2½ inches apart from one another, with the rear stick ½ inch from the back of the frame.

Use a fourth rubber band to fasten the lower ends of both 5-inch sticks together; this will create the handle for the rubber band gun. Adjust if needed.

# Step 4

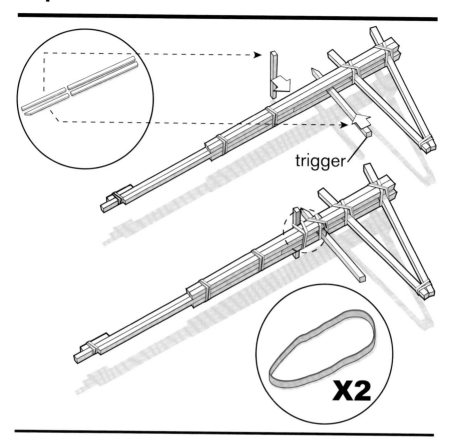

trigger

**X2**

Both 3-inch sticks from chopstick set 1 will be used in this step.

Start with the trigger. Slide the angled stick between the frame, similar to how the handle was constructed. Place the stick roughly ¾ inch in front of the handle, with the angle on top and slanted backward toward the handle. Use one rubber band to hold this stick in place. Because of its middle location, you will have to wrap the rubber band around the frame and trigger. When finished, the trigger can still move back and forth.

With one more rubber band, attach the second 3-inch stick to the top of the rubber band gun, 1 inch in front of the trigger assembly. This stick will be positioned horizontally and will assist in launching the rubber bands.

# Step 5

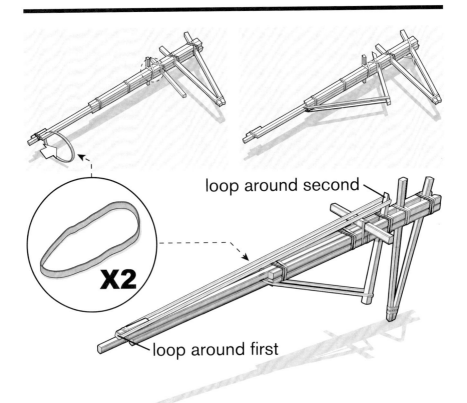

loop around second

X2

loop around first

One of the unique features of the Chopstick Rubber Band Gun is the real working trigger, with recoil. Slide one rubber band between the end of chopstick set 2 until it's positioned against the main frame. Loop the rubber band around the bottom of the trigger so that it pulls the trigger forward when not loaded.

To load the rubber band gun, loop one rubber band around the tip of the barrel first, then stretch and loop the same rubber band around the tip of the trigger. To fire, pull back the bottom of the trigger and watch the rubber band fly. If it doesn't fire, adjust the angle of the trigger.

***Never aim your Chopstick Rubber Band Gun at a human or animal.*** A rubber band can sting or cause eye damage, so ***always wear eye protection when firing.***

# PAPER CLIP GRAPPLING HOOK

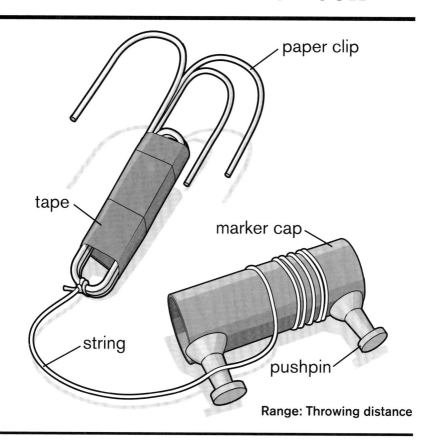

paper clip

tape

marker cap

string

pushpin

**Range: Throwing distance**

The ninja master says, "Trip wires with a grappling hook, not your foot." Avoiding traps is just one of the many uses for the Paper Clip Grappling Hook. Fashioned from paper clips and with an attached rope storage unit, it has all the classic elements of a reliable MiniWeapon tool. With this grappling hook, nothing will be beyond your reach!

## Supplies

3 large paper clips
Electrical tape
2 pushpins (optional)
1 standard art marker cap
   (optional)

## Tools

Safety glasses
Pliers
Thread or dental floss
Scissors (optional)

# Step 1

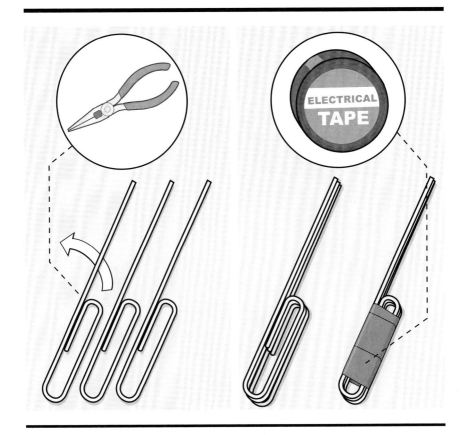

With pliers, straighten out only the large loop on three large paper clips, as shown. Then stack all three paper clips so the silhouettes align. Once stacked, place electrical tape around the assembly, keeping the bottom loops unobstructed.

# Step 2

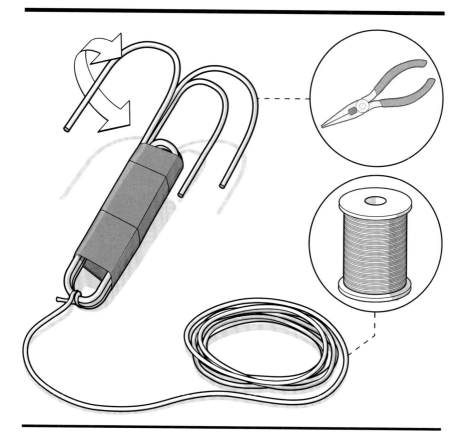

With pliers, bend all three paper clips into grappling hooks, with each of them evenly spaced and arced back toward the main housing. Refer to the illustration for the position of the bends.

Uncoil about 8 feet of thread or dental floss. Tie one end of the thread to the bottom paper clip loops.

# Step 3

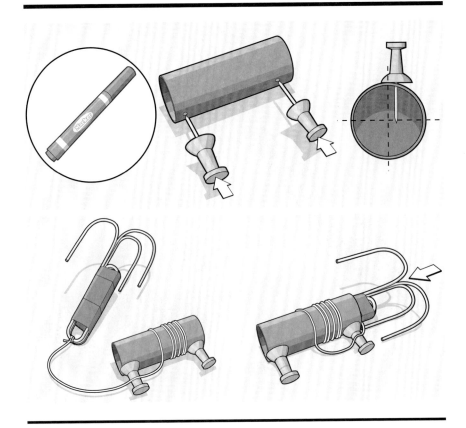

To construct a rope storage unit (optional), attach a pushpin to each end of one standard art marker cap. The pushpins should be off-center so that the paper clip grappling hook frame can still fit snugly inside the marker cap.

Tie the end of the string to the center of the marker cap or to one of the pushpins. Coil the rope around the marker cap and then slide the back end of the grappling hook into the marker cap for storage.

***This grappling hook has pointed paper clip claws, so do not throw it at living targets***.

# HANGER GRAPPLING HOOK

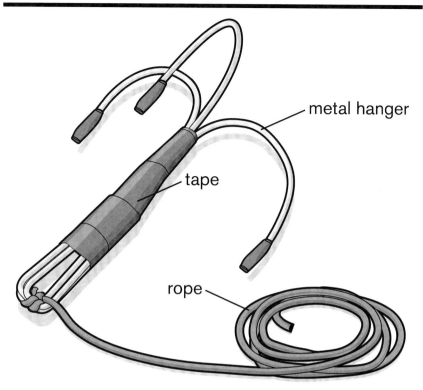

metal hanger

tape

rope

**Range: Throwing distance**

Fabricated from cold steel, the Hanger Grappling Hook is designed for grasping larger objects or tethering rope at unreachable heights. It's small enough to be concealed but big enough for breaching tactical obstacles. With a few select cuts and bends, you'll be scaling walls in no time!

## Supplies

3 metal coat hangers

Electrical tape

1 piece of ⅛-diameter string (8 to 20 feet long)

## Tools

Safety glasses

Wire cutters or pliers

Scissors (optional)

# Step 1

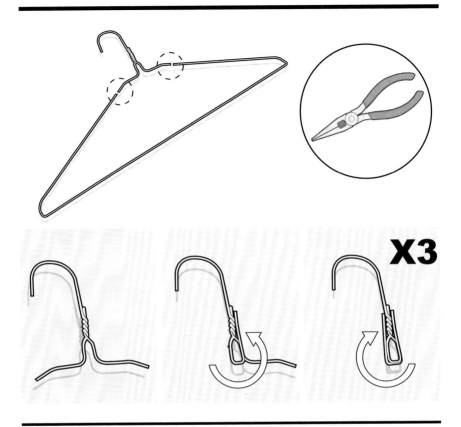

With a pair of wire cutters or pliers, remove the hook end from three metal coat hangers. Cut the hooks off with 2 inches of neck lower frame still attached (top illustration).

Using the pliers, bend the left side of the neck 180 degrees toward the right side. Then, where the bent neck touches the untouched neck, place one 90-degree bend upward. Repeat this bend on the right side by bending the neck 180 degrees to the left side, and then bend it up, opposite the other side.

Repeat this step until all three metal hooks are bent like the lower right illustration.

# Step 2

Stack all three bent hooks with one of the hooks facing the opposite direction. Once stacked, tightly wrap electrical tape around the neck of the assembly. Do not obstruct the bottom loop detail created in step 1.

# Step 3

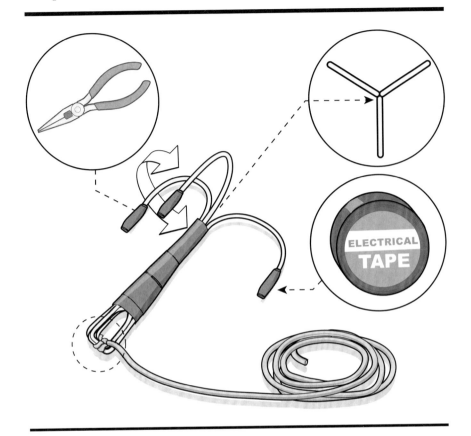

Use the pliers to evenly space the three necks. Then tie 8 to 20 feet of ⅛-diameter string to the bottom loops, cutting the string to length if necessary. To prevent poking or surface scratching, wrap tape around the tips of the grappling hook claws. You can also modify the tips by adding three eraser caps or three removed soft ends from plastic cotton swabs.

The grappling hook is mission ready! ***Be aware that this grappling hook has pointed metal claws, so do not throw it at living targets.***

# 6

# MELEE AND RANGED WEAPONS

# CARDBOARD TUBE KUNAI

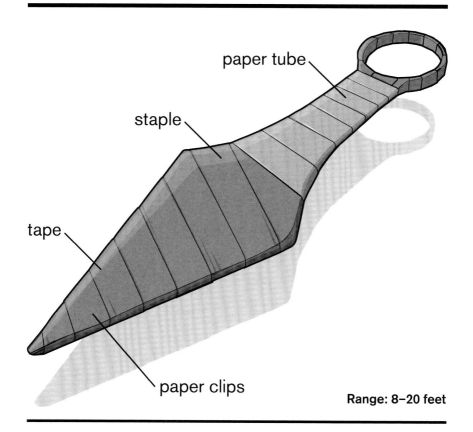

paper tube

staple

tape

paper clips

**Range: 8–20 feet**

A *kunai* combines the art of stealth and the art of deadly craft! Disguised as a worker's tool, no one will suspect its combat potential. Although the Cardboard Tube Kunai is constructed from a few cardboard tubes, its tip is weighted, increasing its ranged weapon potency.

## Supplies

2 toilet paper tubes (or similar)
Black electrical tape
8 paper clips
Red electrical tape (optional)

## Tools

Safety glasses
Scissors
Stapler

# Step 1

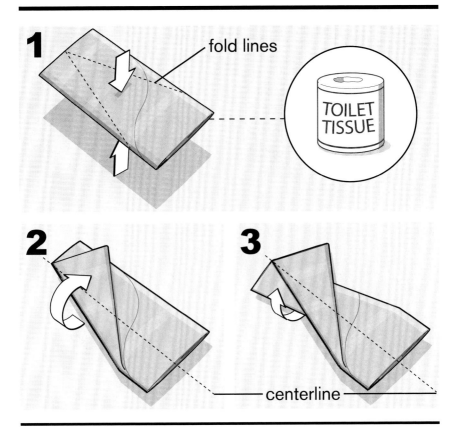

fold lines

TOILET TISSUE

Round up two toilet paper tubes that are 4 inches long or similar. The first tube will be modified for the blade of the kunai. To start, flatten tube 1 as shown, creased on the edges to make a double-walled rectangle (1).

Next, using the center of the tube as the point—the dotted lines on the illustration—fold the left side of the rectangle in a wedge shape, overlapping the right side of the tube (2). The wedge shape should start ¾ inch from the bottom of the compressed tube. Then repeat this fold on the opposite side of the tube by folding the right side of the tube to mirror the left side (3).

# Step 2

Continue folding each flap over one more time to create a point (4 and 5). Secure the point with tightly wrapped electrical tape (6).

# Step 3

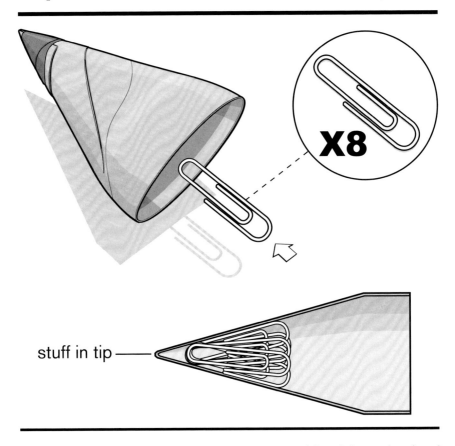

stuff in tip ———

To increase this ranged weapon's accuracy, add weight to the tip of the cardboard blade. Pop open the rear of the cone, then slide eight paper clips into the cavity, wedging all the paper clips inside the point.

# Step 4

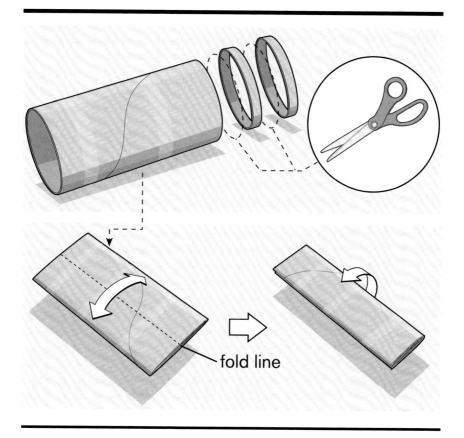

fold line

With scissors, cut two ⅜-inch rings off the second cardboard tube. Set the two rings aside until step 6.

Next, flatten the second cardboard tube into a rectangle. Then fold the compressed tube one time to decrease its width by half.

# Step 5

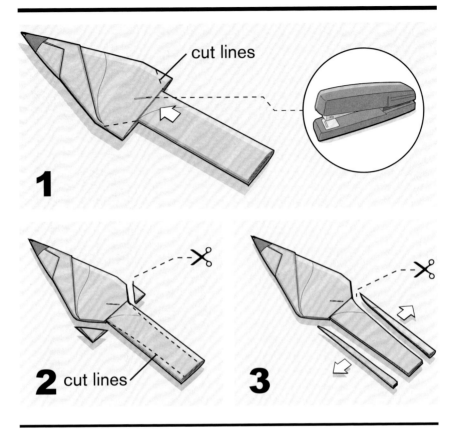

cut lines

**1**

**2** cut lines

**3**

Slide the folded tube into the rear of the cardboard blade, approximately ½ inch, and then staple the folded cardboard tube in place. This will be the kunai handle.

Trim the bottom corners of the cardboard blade to 45 degrees, as shown by the dotted lines (1).

Next, add a slight curve to the length of the handle by trimming each side, also illustrated with dotted lines (2). With scissors, start approximately ³⁄₁₆ inch from the lower edge and taper the cut as you reach the blade (3).

# Step 6

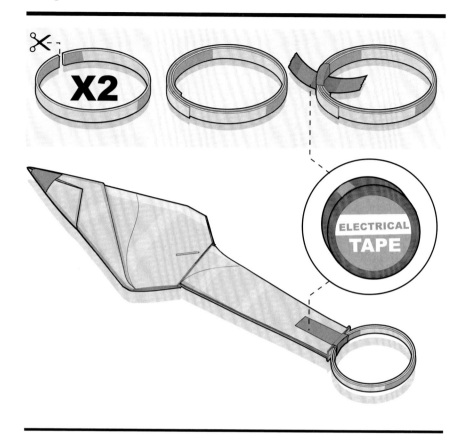

With scissors, cut both ⅜-inch rings from step 4 once. Then reduce the diameter of the ring to approximately 1½ inches. Use the second ring to double up the diameter for strength, then tape the assembly together.

Create a ring mount by folding the lower cardboard handle in opposite directions, ⅛ inch. Then tape the cardboard ring to the folded end.

# Step 7

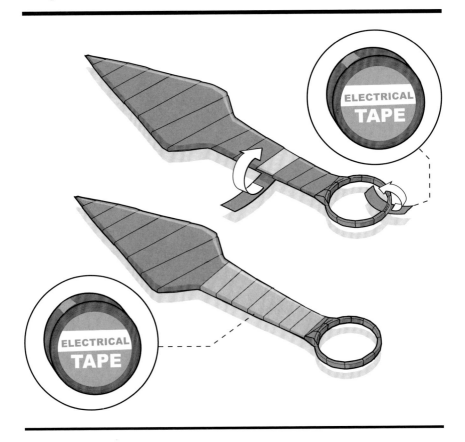

For added strength, wrap the entire cardboard assembly in black electrical tape. Use small sections of tape to cover the ring detail at the end of the handle. To make it look even better, use red electrical tape around the handle, to represent the grip area.

Now you're ready to throw the kunai. Place the blade in your palm, then throw the blade underhand, keeping your wrist straight. ***Never throw the Cardboard Tube Kunai at a living target. Its weighted cardboard point could cause eye damage.***

# PAPER KUNAI

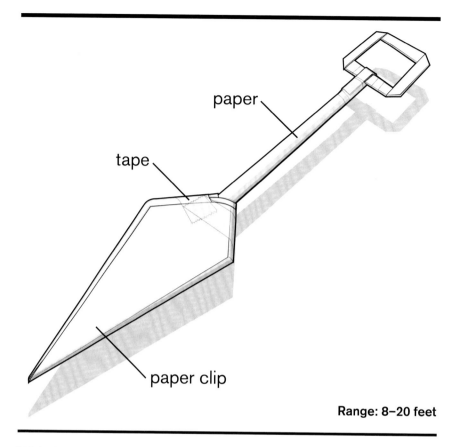

paper

tape

paper clip

**Range: 8–20 feet**

With the help of ninja origami, you can create a Paper Kunai using just two sheets of paper and optional paper clips, costing zero yen! Once constructed, it can be utilized as a close-combat or range-weapon trainer. Plus, this Paper Kunai's simple bill of materials makes it an excellent choice for mass production.

## Supplies

2 sheets of copy paper (8½ inches by 11 inches)
1 small paper clip
Clear tape

## Tools

Safety glasses
Scissors
Pliers

# Step 1

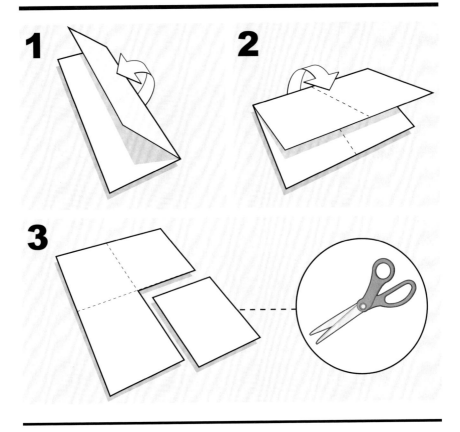

Start by constructing the Paper Kunai blade from one standard sheet of rectangular copy paper. Fold the sheet of paper in half vertically (1). Then fold the sheet of paper again, horizontally (2). Unfold the sheet of paper and use the crease lines as cut lines. Cut the paper into four identical rectangles (3).

# Step 2

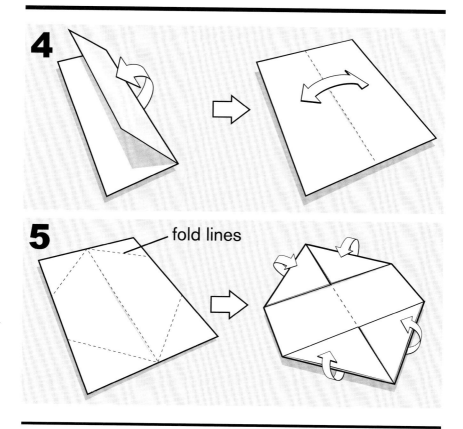

fold lines

Each of the small rectangles will be folded into a wedge. Once all four wedges are complete, they will be assembled into the final blade assembly.

Start by folding one rectangle in half lengthwise to create a crease line down the center of the rectangle (4), then unfold it.

Using the center crease line, fold all four corners inward, lining each of the ends along the crease line. The folded paper should resemble a hexagon, with six sides (5).

# Step 3

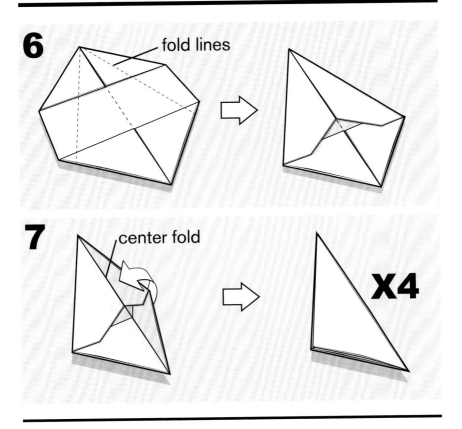

**6** fold lines

**7** center fold

**X4**

Using the dotted lines in the illustration as reference, fold the two sides of the hexagon to create a large wedge, aligning both ends of the fold to the centerline (6).

Fold the wedge in half, with the folds ending up inside the triangle (7). Repeat substeps 4 to 7 three more times until you have four finished wedges.

# Step 4

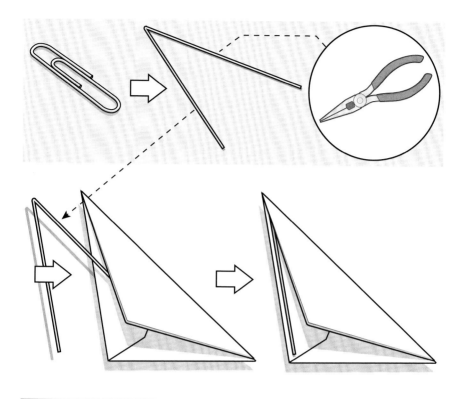

Using pliers, straighten out a small paper clip. Then bend the straight-ened paper clip in half, creating an approximate 22-degree wedge, the same angle as the paper wedge you folded.

Slide the bent paper clip inside the folded paper wedge. You may need to adjust the bend or cut the ends so that the paper clip does not extend outside the wedge.

# Step 5

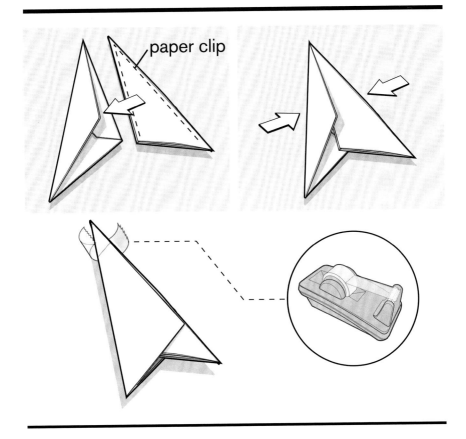

paper clip

Mirror another wedge, opposite from the paper wedge with the stuffed paper clip. Slide the mirrored wedge over the paper clip wedge.

Once both wedge points are aligned, add tape to the tip of the wedge assembly.

# Step 6

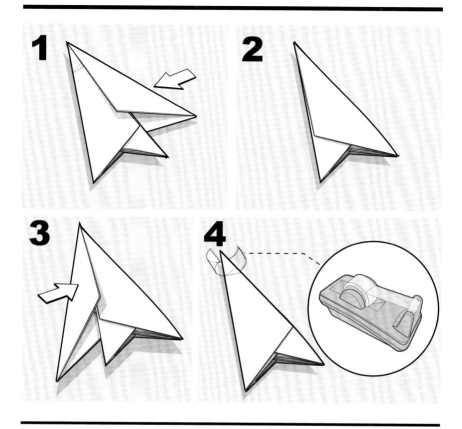

Continue to add the two remaining paper wedges onto the wedge assembly. Slide another wedge over the right side of the wedge, then slide the fourth wedge over the left side of the wedge assembly. Tape the point of the wedge assembly.

# Step 7

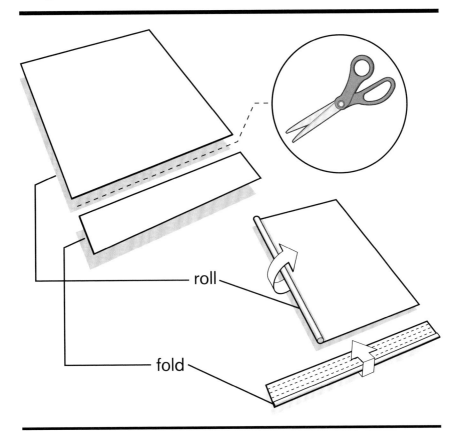

roll

fold

The second sheet of paper will be used for the handle and ring detail of the Paper Kunai. With scissors, reduce the paper's length to 2 inches.

Starting from the sheet of paper's original height, tightly roll the sheet into the handle. Use tape to secure the tube.

The width of the 2-inch strip will be reduced with a series of ¼-inch folds until the final width is only ¼ inches. This strip will create the ring attached to the bottom of the handle.

# Step 8

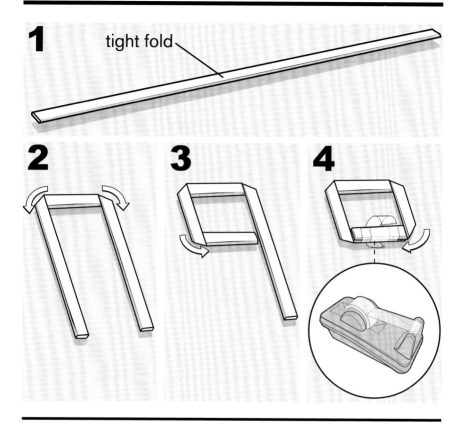

**1** tight fold

**2**

**3**

**4**

To finish the kunai ring detail, fold the ¼-inch strip twice, at opposite ends, with the center at 1¾ inches (2). Continue 1¾ inches down the strip and bend the strip inward to a 90-degree angle (3). Repeat this fold on the opposite end by bending the strip inward and overlapping it to create a "ring." Add tape to hold the ring in place (4).

# Step 9

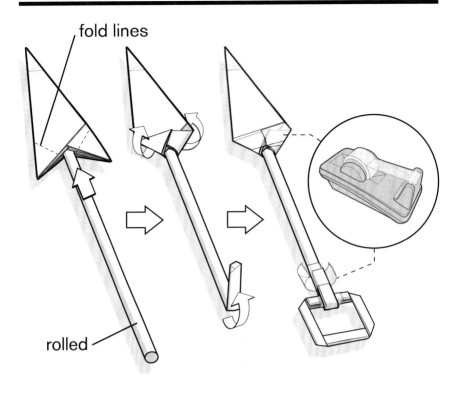

fold lines

rolled

Slide the rolled handle into the back of the paper wedge. With the rolled handle centered, fold the rear points of the blade inward to lock the rolled handle into place, then add tape to the folded area to secure the handle.

Fold the end of the rolled handle 1 inch toward the blade (middle illustration). Place the folded ring inside the handle crease. Add tape to the folded handle to keep the ring in place.

To throw the kunai, place the blade in your palm and then throw the blade underhand, keeping your wrist straight. ***Never throw the Paper Kunai at a living target. Its weighted cardboard point could cause eye damage.***

# CEREAL BOX BOOMERANG

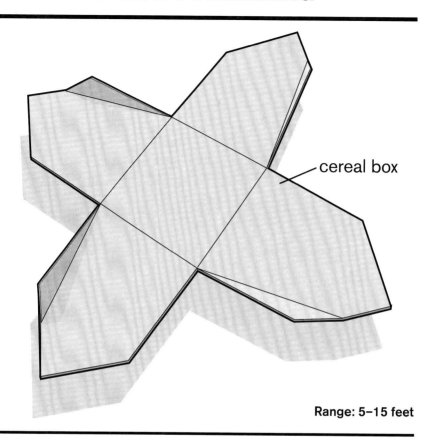

cereal box

**Range: 5–15 feet**

Whether for target practice or entertainment, the Cereal Box Boomerang is a great introduction to the art of the boomerang. It's a quick project, giving a ninja practitioner plenty of time to enjoy its flight! Plus, the large cardboard frame makes it ideal for custom graphics.

## Supplies

1 cereal box

## Tools

Safety glasses
Ruler
Pen
Scissors

# Step 1

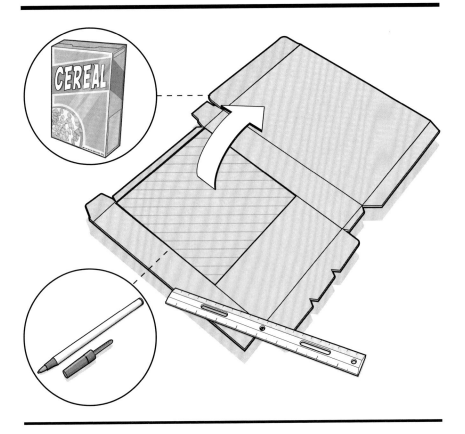

Start with an empty cereal box or cardboard box that is similar in width. Disassemble the box by carefully peeling apart the factory-glued seam that holds the box together.

Use a ruler to measure out a 7½-inch-by-7½-inch square or similar and mark the dimensions with a pen on the blank side (the original inside) of the cereal box.

# Step 2

With scissors, cut the square out of the cardboard. Then, from each corner of the square, mark a 2-inch line on the edge with a pen and ruler. The finished square should have 8 lines along the perimeter as illustrated, each 2 inches from the nearest corner.

# Step 3

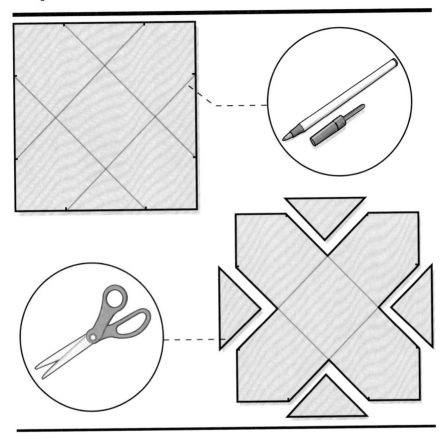

Using the ruler, connect the 2-inch markings, following the top left grid illustration. Notice how each line is angled 45 degrees across the square, connecting to the mark perpendicular to the start line. A total of four lines should be drawn.

Once the guidelines are completed, the lines will create one small triangle, centered, on each side of the square, for four total. With scissors, remove all four of the small triangles.

# Step 4

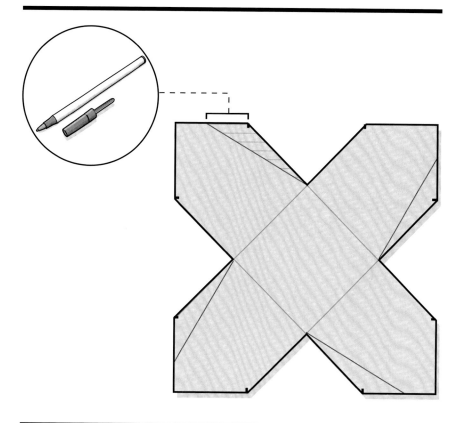

Next, as shown above, add a 1-inch mark on one of the 2-inch sides with a pen. Then, from that 1-inch mark, draw a straight line to the nearest inside triangle. Press the pen firmly to create a fold line.

Repeat this line, moving clockwise, on the coordinating points. When this step is complete, the boomerang should have four added lines converging to the inside points.

# Step 5

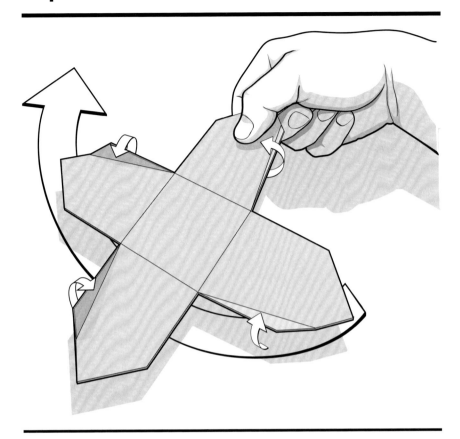

The last step is to fold up all four flaps along the lines drawn in step 4. Each flap should be folded up to a 45-degree angle, which will help the boomerang fly.

Position the Cereal Box Boomerang horizontally (level), holding it at the corner with your thumb on top, as shown. Pull back your arm and bring it forward while flicking your wrist, launching the boomerang at a slight angle upward. Enjoy the flight, but be prepared for its return. **Never throw the Cereal Box Boomerang at a living target. Its thin cardboard points could cause eye damage.**

# ORIGAMI BOOMERANG

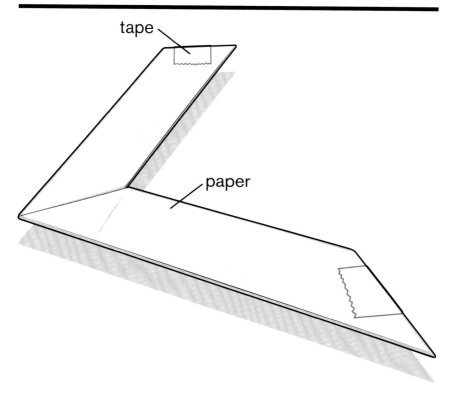

tape

paper

**Range: 5–20 feet**

"In movement you will find balance" with the Origami Boomerang. Cleverly constructed from a single piece of paper, this boomerang always returns to its ninja master. With a single katana cut and a series of choice folds, this MiniWeapon design is indoor friendly and invites experimentation. Enjoy its climbing flight; just be mindful of its return.

## Supplies

1 sheet of copy paper (8½ inches by 11 inches)
Clear tape

## Tools

Safety glasses
Scissors

# Step 1

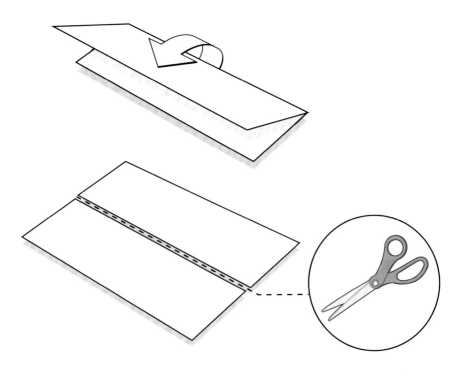

Prepare a standard sheet of copy paper by folding it in half vertically, aligning it edge to edge, and then unfolding it to reveal the crease line down the center.

Using the center crease line as a cutting guide, cut the paper in half with the scissors. The origami boomerang requires only one half of the paper.

The first few steps of the Origami Boomerang are pre-folds prior to the boomerang taking shape; the folds make construction easier in the later steps.

# Step 2

**1** fold

**2** centerline crease

unfold

**3** centerline crease

fold

**4**

fold

With the half sheet of paper positioned vertically, fold the paper once more in half, lengthwise, with the edges aligned (1). Then unfold the paper to reveal the center crease line (2).

Using the center crease line as the go-to line, fold both vertical sides next to the center crease line, for a total of two folds (one per side). Do not overlap the center crease line; leave a bit of space between the two edges (3 and 4).

# Step 3

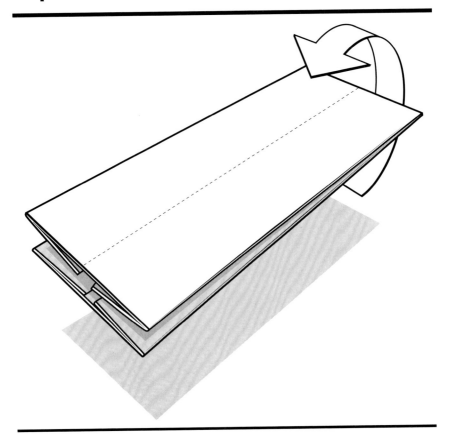

Now fold the paper over end to end, reducing the length by half as shown. When the edges align, run your finger over the new crease line for a crisp fold.

# Step 4

**1** crease end — centerline crease

**2** unfold —

**3** flip over

**4** unfold —

Next, at the crease end, fold both corners over so the top edges are aligned with the center crease line when complete (1). Run your finger over both creases and then unfold both corners (2).

Flip the paper over and repeat the fold on the opposite side. Fold both corners over so the top edges are aligned with the centerline when complete (3). Run your finger over both creases, then unfold both corners again (4).

# Step 5

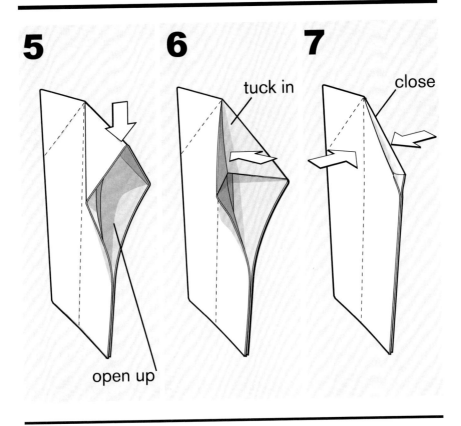

**5**　**6**　**7**

tuck in

close

open up

Open the right side to create a pocket (5). Using the crease lines from the corners, push/tuck the corner into the pocket (6). Then close the flaps, aligning the edges (7).

# Step 6

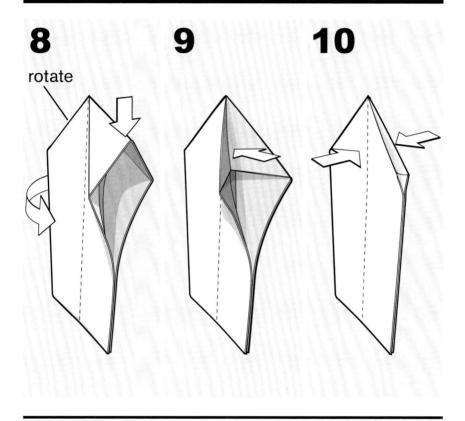

Repeat step 5 on the left side of the paper assembly. Open the left-side pocket (8) and using the crease lines from the corners, push/tuck the corner into the pocket (9). Then close the flaps, aligning the edges (10).

The finished assembly should be a rectangle with one end pointed.

# Step 7

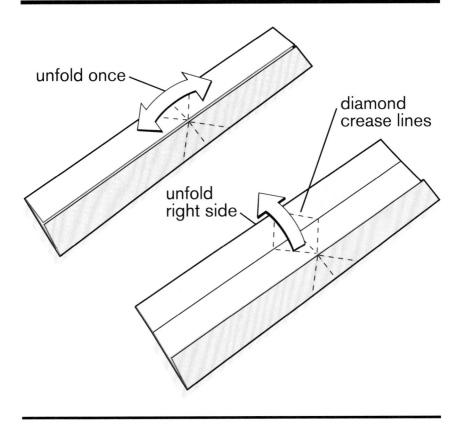

unfold once

diamond
crease lines

unfold
right side

Unfold the paper once (top image) and then unfold the right-side flap only (bottom image). Once open, you will see a diamond crease line, which will assist you in the next fold. You may want to strengthen all four crease lines (dotted lines) by pressing your fingernail along them.

# Step 8

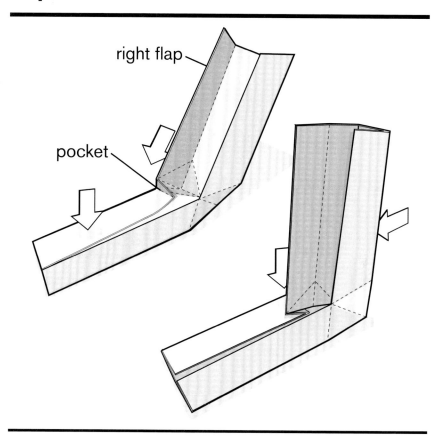

right flap

pocket

Laying the paper down, close the bottom half of the right flap, creating an overlapping pocket. When you do this, the left side will pop up.

Continue folding the paper by flattening the center pocket, with the top right flap positioned at a 90-degree angle.

# Step 9

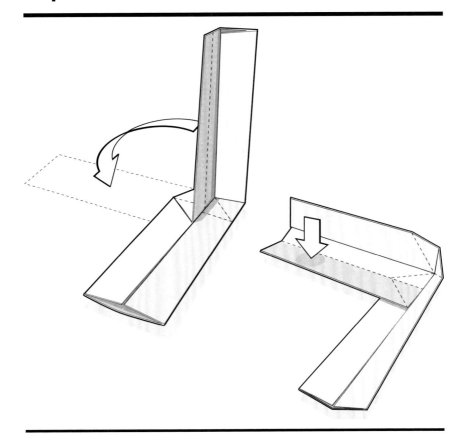

Now fold the upright 90-degree flap down, so that the paper *L* is on the same plane.

# Step 10

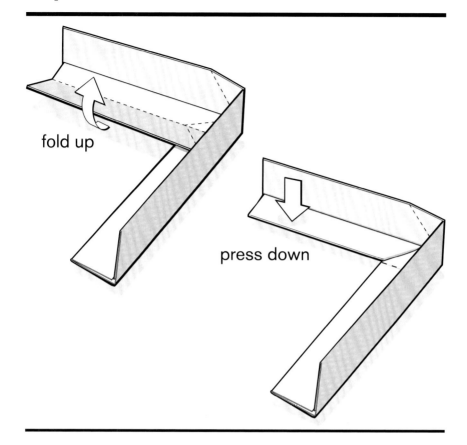

fold up

press down

Fold the opened right flap back into the closed position, toward the midline. Use your fingernail to press the crease line flat. The Origami Boomerang is starting to take shape.

# Step 11

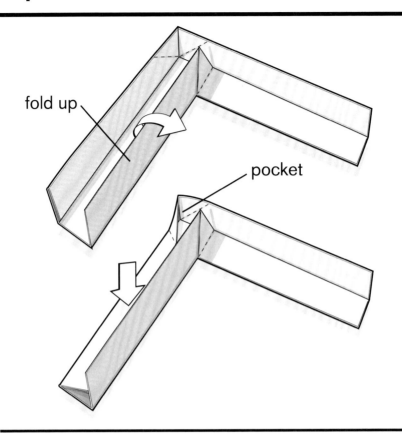

fold up

pocket

Now position the paper like the illustration above. Lift up the inside left panel (top image). Then fold the outer left-side panel down, creating a pocket in the center (bottom image).

# Step 12

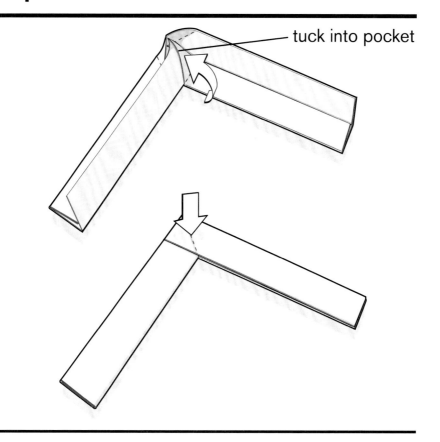

tuck into pocket

Next, fold the inside left panel over the outside panel. Work the inside corner into the center pocket, tucking it in place (top image). Press the flap down and run your fingernail along the crease edges to flatten out the boomerang (bottom image).

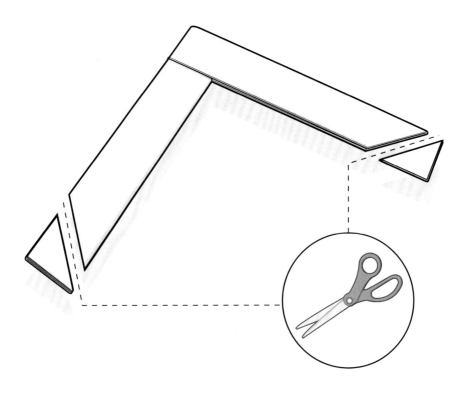

A true Origami Boomerang requires several more steps to lock the paper ends into place. However, a ninja doesn't have time for that!

With scissors, cut two corresponding 45-degree angles at each end of the boomerang. Compare your boomerang to the image above prior to cutting.

# Step 14

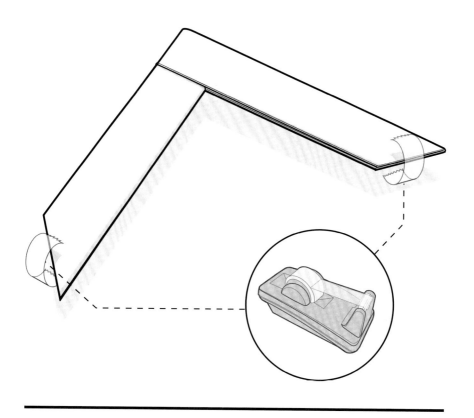

Use a small amount of clear tape to hold the end flaps together.

# Step 15

The Origami Boomerang is now finished! Position the Origami Boomerang horizontally, holding the diagonal-line side upward with your thumb on top as illustrated. Pull back your arm and bring it forward while flicking your wrist, launching the boomerang at a slight angle upward. Enjoy the flight, but be prepared for its return. ***Never throw the Origami Boomerang at a living target. Its thin paper points could cause eye damage.***

# MAGAZINE NUNCHUCKS

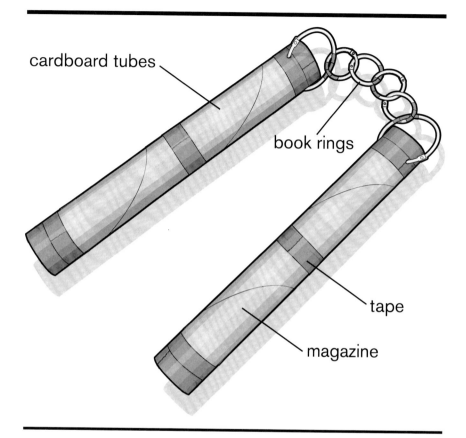

cardboard tubes

book rings

tape

magazine

The Magazine Nunchucks, also known around the dojo as *nunchaku* or chainsticks, are an excellent training weapon for a ninja balancing mind and body. Constructed from two tightly rolled magazines and a home-made metal chain, they are seriously durable. With a few nunchuck swings, you'll be yelling "Cowabunga!"

## Supplies

2 magazines

2 paper towel tubes (more than 10½ inches long)

2 large metal book rings (2-inch diameter)

4 small metal book rings (1¼-inch diameter)

Electrical tape

## Tools

Safety glasses

Scissors

Power drill with ¼-inch drill bit (or similar)

# Step 1

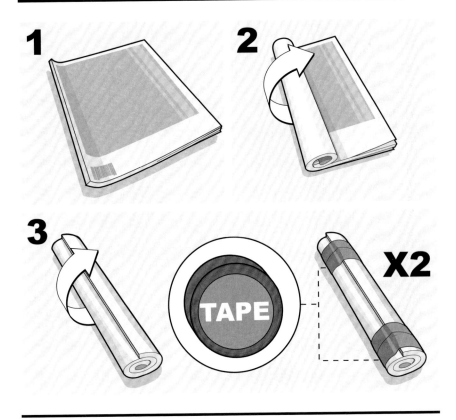

Start with two old magazines, similar in size and page count (1). Tightly roll each magazine horizontally (2), then use electrical tape to secure both ends of the cylinder (3). Each magazine cylinder should have a diameter of approximately 1 inch.

# Step 2

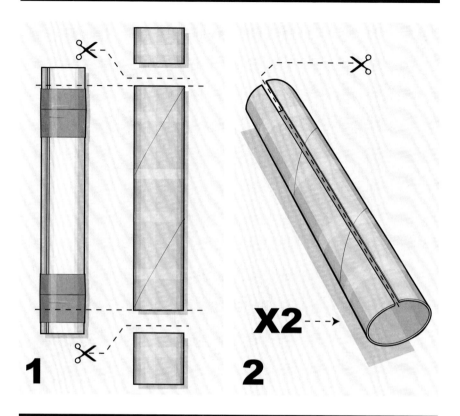

**1**

**X2**  →

**2**

Paper towel cardboard tubes vary in size, so before you begin construction, measure to confirm that the length of the tube is similar to the rolled magazines' length.

With scissors, cut both cardboard tubes ½ inch shorter than the length of the rolled magazine tube. Discard the scrap ends (1).

With the tip of the scissors inside the cardboard, make one continuous cut along the length of the cardboard tube as shown (2). Repeat this cut on the second cardboard tube.

# Step 3

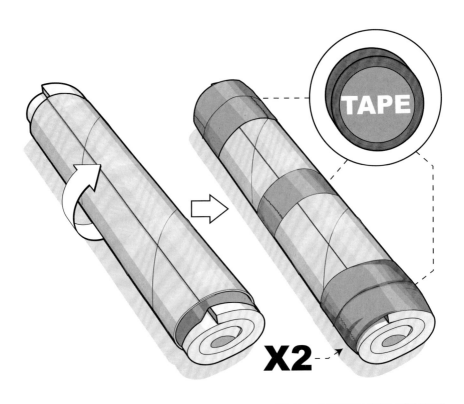

Wrap the modified cardboard tube around the rolled-up magazine, with the magazine protruding from both ends. Place electrical tape at both ends and in the middle to fasten the tube in place. Repeat this step so both magazine "sticks" are similar.

# Step 4

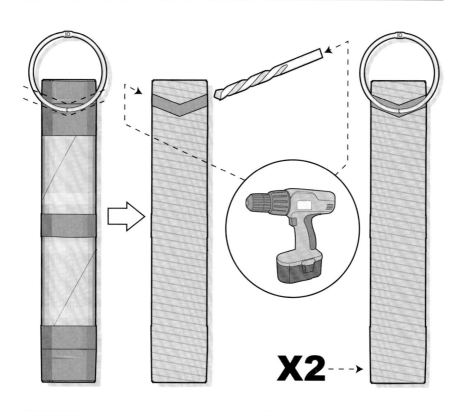

**X2** - - - →

To fasten the chain onto the two magazine sticks, a 2-inch metal book ring will be mounted to one end of each of the rolled-up magazines. A power drill with a ¼-inch drill bit will be used in this step. ***Take special care to avoid injury; young ninjas should get an adult's help.***

Starting ½ inch from the end of the roll, drill a ¾-inch deep hole at about a 30-degree angle, as shown, toward the center of the rolled magazine. Opposite that hole, drill an identical hole on the reverse side of the cardboard tube, connecting the hole in the center of the tube.

Rotate the 2-inch book ring into the hole as shown in the right illustration. You may need to do some additional drilling if the book ring is obstructed.

# Step 5

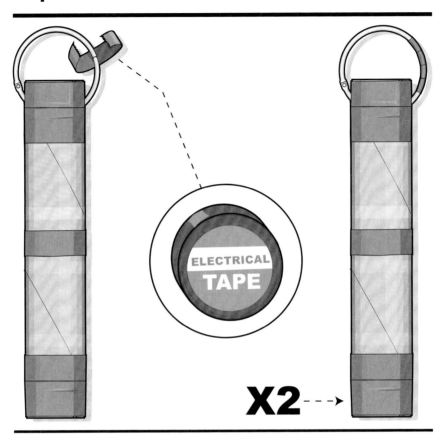

**X2** - - - ▸

Tightly tape the book ring clasp shut—using electrical tape is especially important in this step because of its width and durability. Place additional tape around the clasp to ensure the connection is secure. Then rotate the covered clasp to one side or the other, but not directly on top.

# Step 6

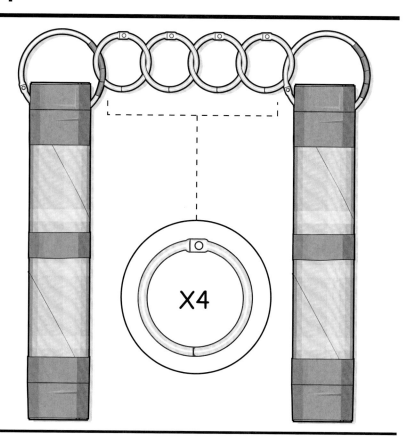

Connect both magazine sticks by clasping four 1¼-inch metal book rings to each of the large attached rings.

# Step 7

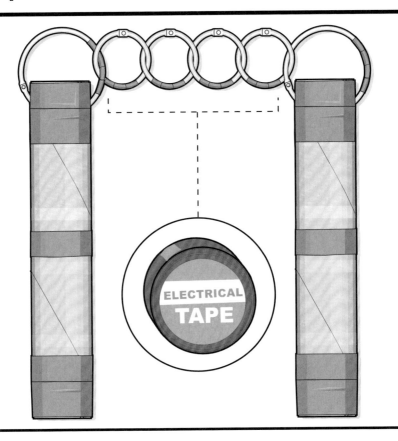

Tightly tape each book ring clasp shut, using several small pieces of tape around the clasp area. It is important that the added tape is properly applied, because this tape will prevent the clasp from flinging open when operating the Magazine Nunchucks, a malfunction that could cause injury.

Time to practice! Holding one handle, start slowly by twirling the second stick away from your body. Once you feel confident with the simple twirl, you can attempt a figure eight motion with the sticks. Additional combat moves include trying to pass the sticks behind your back or from side to side. **Be sure to start off slowly, and never swing the nunchucks in an uncontrolled manner.** They can cause harm if used improperly. **MiniWeapons projects are not meant for living targets.** Always stay clear of spectators when practicing with the nunchucks. Homemade weaponry can malfunction.

# CARDBOARD HANGER SAI

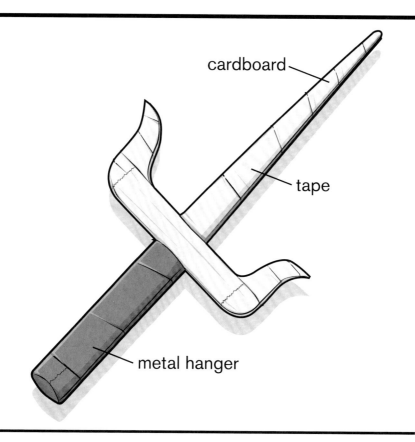

cardboard

tape

metal hanger

You don't flirt with the enemy, you take the enemy down with this Cardboard Hanger Sai! The *sai* is a melee (hand-to-hand) combat weapon most effective for trapping and blocking attacks at close range. Sai are most effective in pairs, so make multiples. This design is easy to replicate.

## Supplies

2 sheets of copy paper (8½ inches by 11 inches)

Clear tape

1 piece of corrugated cardboard (20 inches by 16 inches)

1 metal coat hanger

Black electrical tape (or similar)

Metallic duct tape (or similar)

## Tools

Safety glasses

Ruler

Marker

Large scissors

Pliers or wire cutters

# Step 1

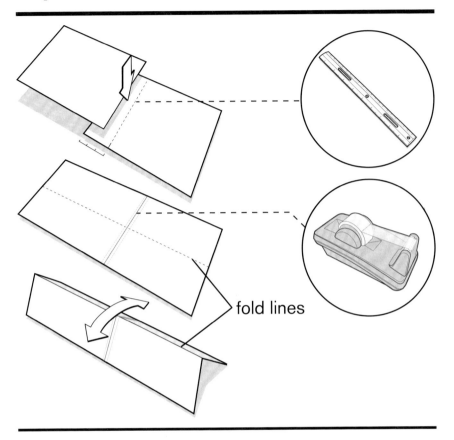

fold lines

To make construction of the Cardboard Hanger Sai easy, a paper template will be used. You can reuse this template to make a second sai if materials are available.

Start with two standard sheets of copy paper. Using a ruler and marker, mark 3 inches from the top of one sheet of paper. Then overlap the two sheets of paper, aligning the edges on the mark as shown, for a combined length of 19 inches. Use clear tape to secure the two sheets as shown in the center illustration.

Next, divide the width of the combined paper in half by folding the center.

# Step 2

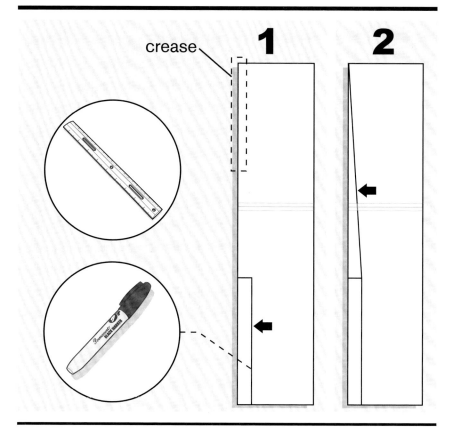

crease 1 2

With the paper still folded, position the paper vertically with the crease running along the left side. Use a ruler and marker to draw a 7-inch by ¾-inch box vertically along the lower left corner (1). This is the start of the sai handle.

Next, draw a roughly 12-inch line connecting the top left corner of the paper to the upper right box corner, indicated by the black arrow (2).

# Step 3

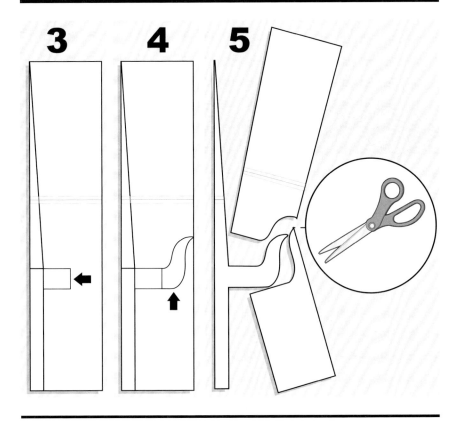

Draw another box off the edge of the 7-inch by ¾-inch vertical box. This box will be 2½ inches by 1 inch, extending out and indicated by the black arrow (3).

Draw the sai prong detail following the pattern (4), with a height of around 3½ inches.

Use scissors to remove the paper around the completed sai drawing (5). Both the top right and bottom right pieces can be discarded or recycled.

# Step 4

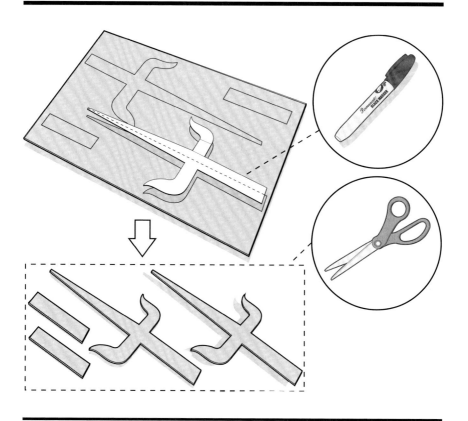

Unfold the paper template and place it on a 20-inch-by-16-inch piece of corrugated cardboard. Position the template on the cardboard so that there is adequate room for two tracings. A single sai will require two pieces of cardboard. Trace two sai outlines on the cardboard with a marker as shown.

Use the template once more to trace two handle rectangles on the cardboard. These added rectangles will be used to strengthen the handle area.

Once the tracing is done, cut out all four elements using large scissors. When finished, you should have two cardboard sai and two cardboard handles (rectangles).

# Step 5

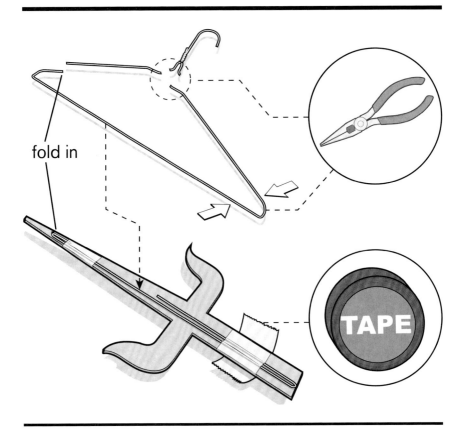

fold in

TAPE

The sai frame will be supported by a modified metal coat hanger. Using pliers or wire cutters, cut off the hook detail from the coat hanger. Discard the hook. Then make an additional cut 1 inch up from the lower corner. Use the pliers to bend in the 1-inch segment tightly against the long segment. This will ensure the tip is rounded rather than sharp in case the sai malfunctions during training.

Compress the large angled section of coat hanger, narrowing its width, then position the bent coat hanger on top of one of the cardboard sai, with the doubled-up wire end on top of the handle end. Use electrical tape or duct tape to secure the wire in place.

# Step 6

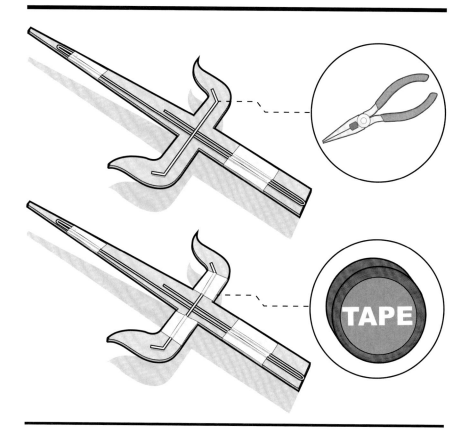

The remaining small section of the hanger will be positioned horizontally, to support both sai prongs. With pliers or wire cutters, bend both ends of the wire up, to fit within the design of the sai prongs. Then use electrical tape or duct tape to secure the metal bar in place.

# Step 7

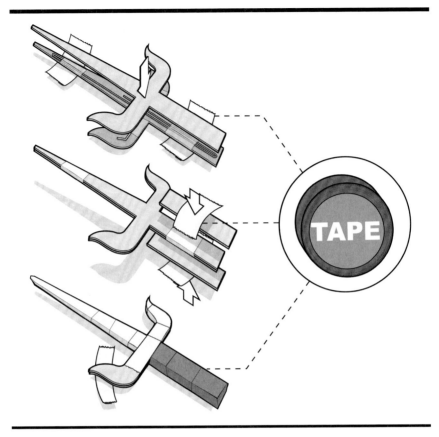

Sandwich the first attached wire by placing the second cardboard sai outline on top. Align the edges and tape the cardboard in place (top images) with electrical tape or duct tape. Add the two cardboard rectangles to the handle area. Attach one per side, to increase the handle width. Use tape to hold them in place.

Wrap the entire sai frame and both sai prongs with tape—metallic duct tape is recommended. You can add black electrical tape to the handle to distinguish the gripping area for added realism.

You've completed the Cardboard Hanger Sai! **MiniWeapons projects are not meant for living targets.** Always stay clear of spectators when practicing with the sai and operate it in a controlled manner.

# CRAFT STICK KATANA

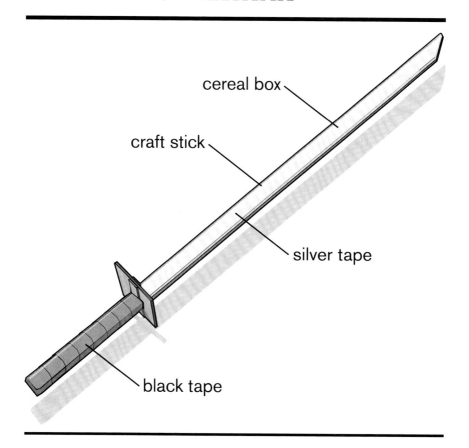

cereal box

craft stick

silver tape

black tape

Behold the Craft Stick Katana, the sword of the ninja! Handcrafted specifically for each ninja, the Craft Stick Katana is an incredible project that transforms a handful of craft sticks and a cereal box into a realistic role-playing marvel. With a metallic-coated blade and a black-wrapped handle, it's one of the coolest MiniWeapons to come out of the lab!

## Supplies

1 cereal box (17 ounces)
White glue
32 large craft sticks
Black electrical tape (or similar)
Metallic duct tape (or similar)

## Tools

Safety glasses
Ruler
Pen
Scissors
Hot glue gun
Hobby knife

# Step 1

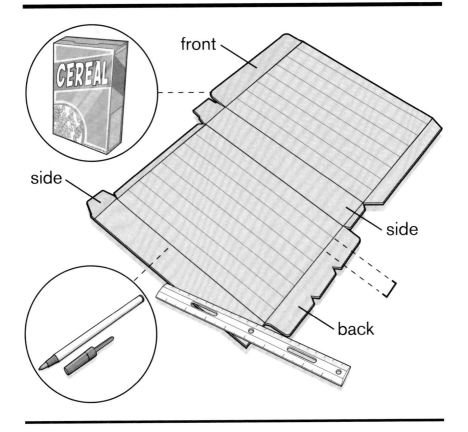

Start with a large, empty 17-ounce cereal box or similar cardboard box. Disassemble the box by carefully peeling apart the factory-glued tabs that hold the box together. Open the box up, with the printed (outside) side down.

With a ruler and pen, draw a grid on the front and back panel area with 1-inch vertical strips as shown on the drawing. Make as many 1-inch strips as can fit inside the front and back panels, but do not grid the side panels.

# Step 2

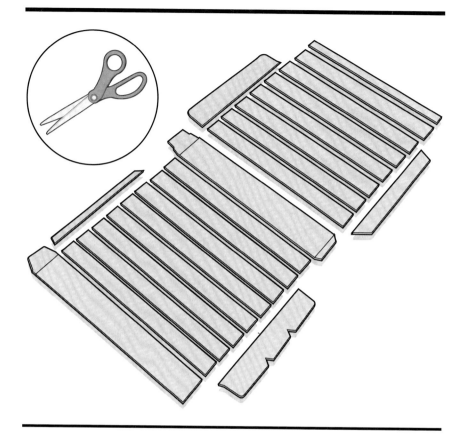

With scissors, follow the pen lines and cut out all the 1-inch strips from the box. Do not discard the box sides; they will be used for the katana hilt guard, also known by its proper name, *tsuba*.

# Step 3

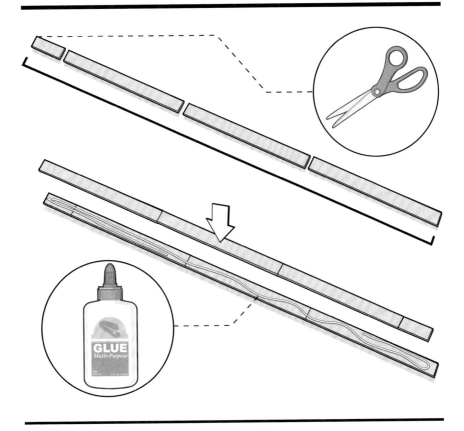

The core of the Craft Stick Katana will be made from three layers of laminated, 1-inch cereal box strips. With a ruler and scissors, position a few strips end to end, for a total length of 38 inches.

Then stagger a second row of strips on top of the first, using white glue to glue the strips in place. Staggering the strips will increase the blade's strength, so don't let the connection points align with the previous layer's. Instead, offset the new strips. Use a ruler and scissors to cut the total strip assembly to a length of 38 inches.

# Step 4

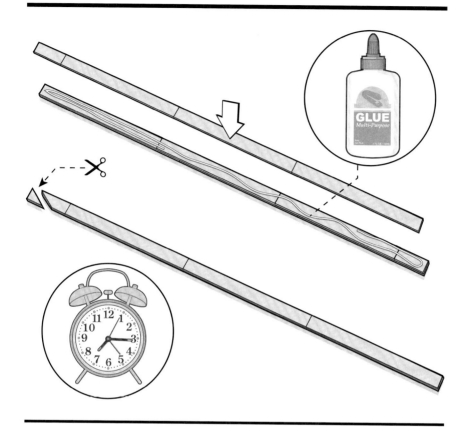

Add the third and final layer of staggered strips on top of the assembly. Use white glue to fasten it in place, and a ruler and scissors to cut the overall assembly to a 38-inch length.

With scissors, cut a 45-degree angle at one end of the strips. Let the glue dry.

# Step 5

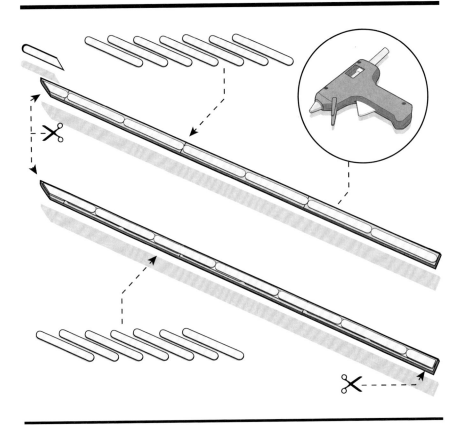

Hot glue seven large craft sticks (or as many as it takes) from end to end along the laminated cardboard strip, positioned along the center. Use scissors at the angled tip to cut a 45-degree angle in the last craft stick (top).

On the same side, stagger another row of craft sticks on top of the first row. It is important that these craft sticks are offset for added strength.

# Step 6

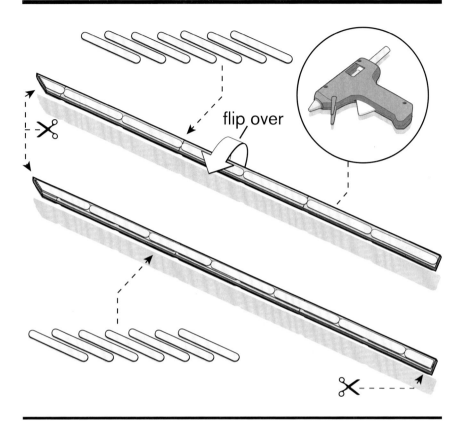

flip over

Flip the assembly over and repeat step 5 on the reverse side of the katana. Start by hot gluing seven large craft sticks from end to end along the laminated cardboard strip, positioned along the center. Use scissors at the angled tip to cut a 45-degree angle in the last craft stick (top).

Again, on the same side, stagger another row of craft sticks on top of the first row. Remember to offset these craft sticks for added strength.

# Step 7

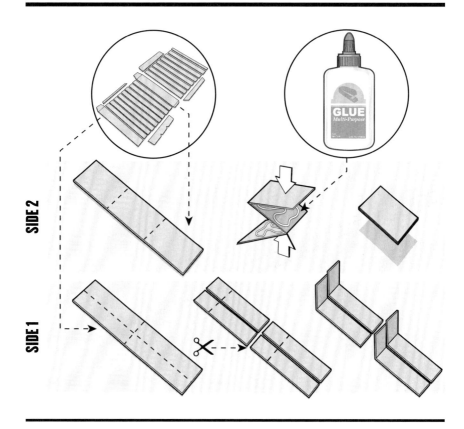

SIDE 2

SIDE 1

The sides of the cereal box will be transformed into the katana hilt guard. Box sizes do vary, but each side should be approximately 2½ inches by 11 inches.

Start with one of the removed cereal box sides (side 1). Accordion-fold the rectangle twice as shown, dividing the length by three. Then apply hot glue between the folds to secure the assembly.

With scissors, cut the second cereal box side into four smaller rectangles (side 2). Then add a 90-degree fold to each small rectangle 2 inches from the end. These four smaller rectangles will be used for mounting the katana guard.

# Step 8

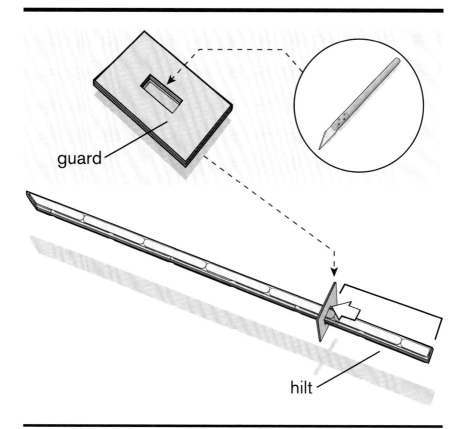

guard

hilt

With the hilt width as reference, use a hobby knife to cut a similar sized rectangle in the center of the tripled-stacked cardboard guard from step 7.

Slide the guard over the hilt and up the blade, resting 8¼ inches from the end.

# Step 9

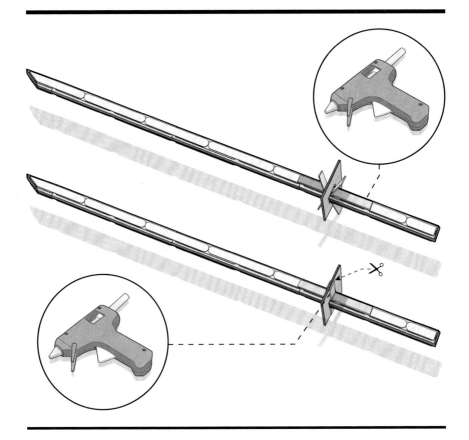

Hot glue all four rectangle mounts to the sword, two on the blade side of the guard and two on the hilt side of the guard, with the 90-degree bends snug to the base of the loose guard.

Once in place, hot glue the four 90-degree bends to the top and bottom surface of the guard. When the glue is dry, use scissors to remove any material that overhangs the cardboard guard.

# Step 10

To increase the diameter of the hilt (handle), add one more layer of craft sticks to both sides. Use scissors to trim the craft sticks flush with the "pommel" (bottom of handle).

Wrap black electrical tape around the hilt for the cord wrap. This will provide adequate gripping when the blade is swung.

# Step 11

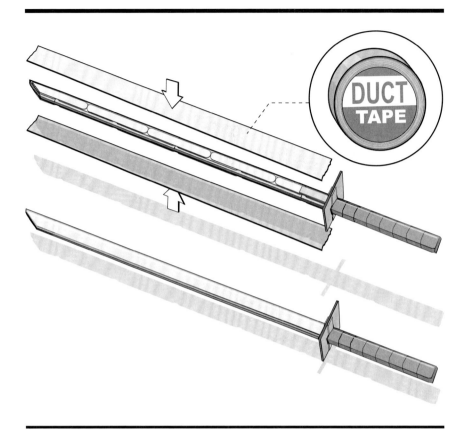

Carefully lay down one large piece of metallic duct tape atop the blade length as illustrated. Wrap the edges of the tape around the blade and the 45-degree sword tip. Then flip the blade over and lay another long piece of duct tape across the opposite edge of the sword, again wrapping the edges around the blade.

The Craft Stick Katana is finished! The sword is durable; however, extreme training could snap the sticks. ***MiniWeapons projects are not meant for living targets.*** Always stay clear of spectators when practicing with the katana, and operate the sword in a controlled manner.

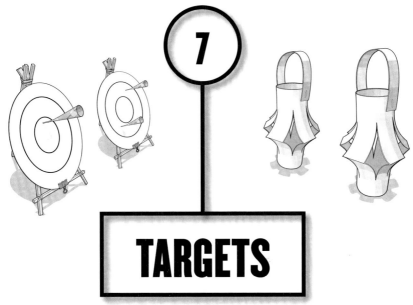

# 7

# TARGETS

# CHOPSTICK TRIPOD

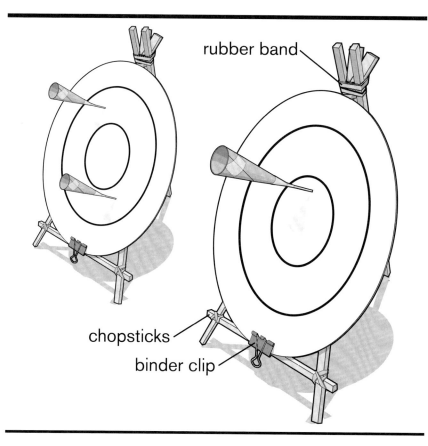

rubber band

chopsticks

binder clip

Earn respect from your fellow ninjas by displaying your MiniWeapon accuracy with the Chopstick Tripod. With a quick assembly and an authentic stance, the tripod works best with blowgun and throwing darts projects. If available materials allow, build multiple targets and set up at different ranges to become a true master.

## Supplies

3 unseparated sets of disposable chopsticks
4 rubber bands
1 disposable paper plate
1 small binder clip (19 mm)

## Tools

Large scissors
Marker

# Step 1

Start with three sets of chopsticks. Snap two sets of chopsticks apart, making four singles and one double. Then, with large scissors, cut one of the single chopsticks in half (1).

Use two of the standard-length chopsticks and one halved section to construct a triangular frame, securing the connections with rubber bands (2).

To hold up the tripod, rubber band the remaining double (attached) set of chopsticks around the top of the triangle, with the lower end supporting the tripod (3). If you snap the set apart, just use a rubber band to secure it.

With scissors, cut a paper plate to a diameter of approximately 6 inches. (If a paper plate is unavailable, copy paper, cardboard, or Styrofoam can be substituted.) Optionally, you can use a marker to draw target rings and numbers onto the face of the plate (4). Then attach the paper plate to the tripod with a small binder clip or tape (5).

# LANTERN KNOCKOUT

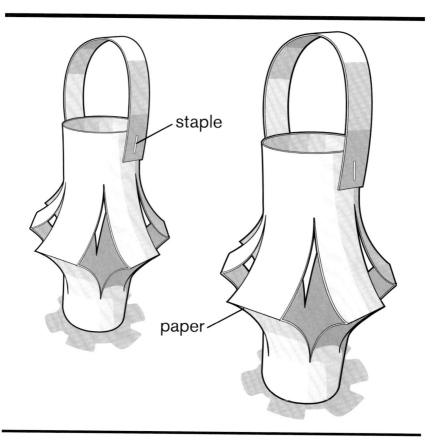

staple

paper

Infiltrating under the cloak of darkness could be the key to a success-ful ninja mission! Practice the art of stealth with the Lantern Knockout, an easily constructed paper target that is designed for throwing star practice. On the ground or hanging from a string, this is one target you don't want to miss!

## Supplies

1 sheet of copy paper (8½ inches by 11 inches)

## Tools

Scissors
Ruler
Stapler

# Step 1

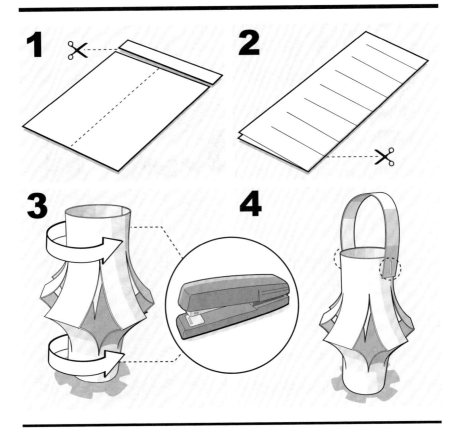

With scissors, cut 1 inch off the short edge of the sheet of paper (1). Then fold the paper in half lengthwise with the edges lined up. Use scissors to make several 3-inch-long cuts spaced 1½ inches apart (2).

Unfold the paper. Position it so the cuts are vertical and roll it to connect the sides; the resulting tube should have a diameter around 2¾ inches. With a stapler, secure the top and bottom of the assembly (3). You may want to press the lantern down so the angled cuts pop out.

The removed 1-inch strip will be the lantern's handle. Staple one end to the top of the lantern, loop the strip to the opposite end, and then secure the end with staples (4).

# EVIL NINJA TARGETS

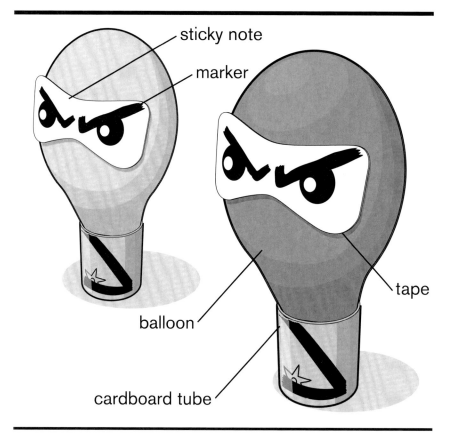

sticky note

marker

tape

balloon

cardboard tube

Constructed from a small balloon and a cardboard tube, an Evil Ninja Target is a perfect way to target practice with your homemade Mini-Weapon blowgun arsenal. Scatter the targets around the training room and unleash balloon-popping fury!

## Supplies

1+ square sticky note (3 inches by 3 inches)

1+ small balloon

Tape (any kind)

1+ toilet paper tube (or similar)

## Tools

Marker

Scissors

# Step 1

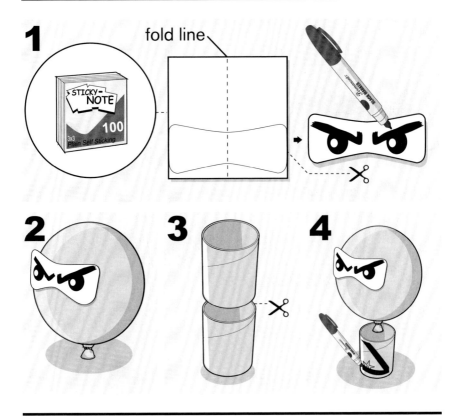

Start by constructing a ninja mask. Fold one 3-inch-by-3-inch sticky note in half. With a marker, draw a half-mask pattern on one half of the sticky note, as shown. The center of the mask (between the eyes) should be located at the crease. With scissors, cut out the drawn pattern. Then use a marker to add evil ninja eyes (1).

Blow up a small balloon. Attach the ninja mask to the front of the balloon with tape (2).

The target's "body" will be constructed from a toilet paper tube. With scissors, cut one toilet paper tube in half to reduce the length to approximately 2 inches (3). Attach the balloon ninja head to the top of the cardboard tube by gently wedging the balloon in place, or use tape. One it's in place, add optional ninja graphics to the tube (4).

# ANCIENT CUP SERPENT

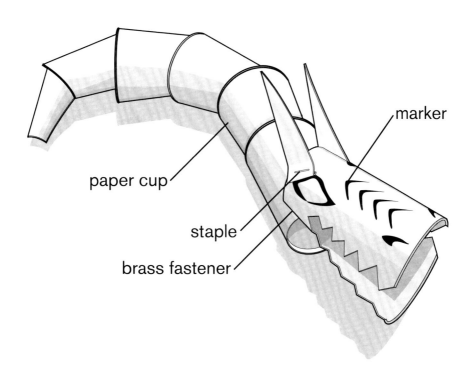

marker

paper cup

staple

brass fastener

Dragons are no myth—these legendary serpents can be summoned! Most are called to protect ancient treasures, but some serpents are sent beyond the realm for grimmer assignments. Don't get caught off guard—practice with the Ancient Cup Serpent, a unique build that transforms paper cups into menacing creatures!

## Supplies

6+ disposable paper cups
6+ brass fasteners
Tape (any kind, optional)

## Tools

Scissors
Stapler
Marker

# Step 1

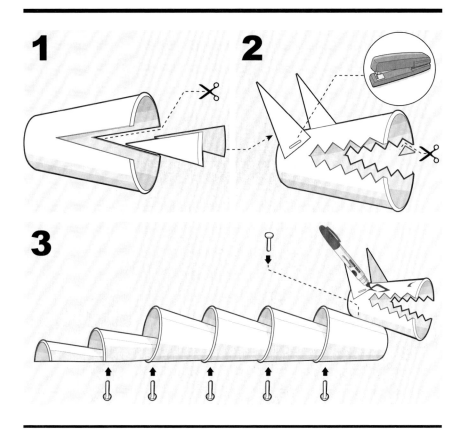

With scissors, cut two wedge sections out of the top of one paper cup as shown. The wedges should be across from each other and cut ¾ of the distance into the cup (1).

Staple the two removed wedges to the back of the cup, point up, to create two serpent horns. Add teeth by cutting zigzag patterns up and down the wedge detail (2).

Attach this "head" cup on top of the back of another paper cup using a brass fastener. Tape can be added to the brass fastener to secure it in place. Once the head is attached, add more cups to the body of the serpent using brass fasteners. When you reach the tail, cut one of the cups in half so that the tail can taper off. When you're finished, use a marker to draw serpent eyes on the head cup (3). In addition, you can mark point values on the individual cup segments to add a scoring element.

# NINJA HIDEOUT

Competitor_____ Date_____

Competitor Signature_____

*Use a copy machine to make multiples and enlarge.*

# BLOWGUN TARGET

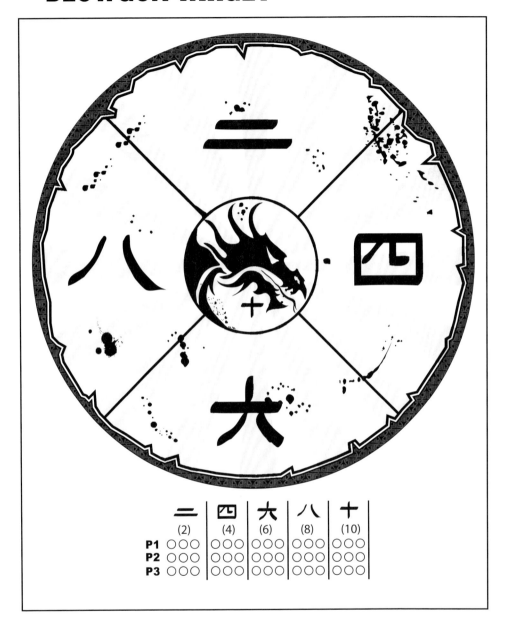

Competitor_____ Date_____

Competitor Signature_____

*Use a copy machine to make multiples and enlarge.*

# WATERMELON SLICE

Competitor_____ Date_____

Competitor Signature_____

*Use a copy machine to make multiples and enlarge.*

For more information and free
downloadable targets, please visit:

# MINIWEAPONSBOOK.COM

*DON'T FORGET TO JOIN THE
MINIWEAPONS ARMY ON FACEBOOK:*

Miniweapons of Mass Destruction:
Homemade Weapons Page

# ALSO FROM CHICAGO REVIEW PRESS

## MiniWeapons of Mass Destruction: Build Implements of Spitball Warfare

John Austin

978-1-55652-953-5
$16.95 (CAN $18.95)

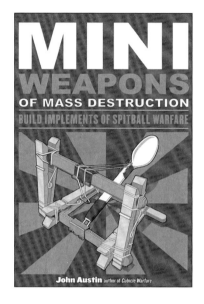

We've come a long way from the Peashooter Era! Using items that can be found in the modern junk drawer, troublemakers of all stripes have the components they need to assemble an impressive arsenal of miniaturized weaponry.

*MiniWeapons of Mass Destruction: Build Implements of Spitball Warfare* provides fully illustrated step-by-step instructions for building 35 projects, including:

- ➲ Clothespin Catapult
- ➲ Matchbox Bomb
- ➲ Shoelace Darts
- ➲ Paper-Clip Trebuchet
- ➲ Tube Launcher
- ➲ Clip Crossbow
- ➲ Coin Shooter
- ➲ Hanger Slingshot
- ➲ Ping-Pong Zooka
- ➲ And more!

And for those who are more MacGyver than marksman, the author also includes target designs, from aliens to zombies, for practice in defending their personal space.

# MiniWeapons of Mass Destruction: Build a Secret Agent Arsenal

John Austin

978-1-56976-716-0
$16.95 (CAN $18.95)

If you're a budding spy, what better way to conceal your clandestine activities than to miniaturize your secret agent arsenal? *MiniWeapons of Mass Destruction: Build a Secret Agent Arsenal* provides fully illustrated step-by-step instructions for building 30 different spy weapons and surveillance tools, including:

- Paper Dart Watch
- Rubber Band Derringer
- Pushpin Dart
- Toothpaste Periscope
- Bionic Ear
- Pen Blowgun
- Mint Tin Catapult
- Cotton Swab .38 Special
- Paper Throwing Star
- And more!

Once you've assembled your weaponry, the author provides a number of ideas on how to hide your stash—inside a deck of cards, a false-bottom soda bottle, or a cereal box briefcase—and targets for practicing your spycraft, including a flip-down firing range, a fake security camera, and sharks with laser beams.

# MiniWeapons of Mass Destruction: Build Siege Weapons of the Dark Ages

John Austin

978-1-61374-548-9
$16.95 (CAN $18.95)

In a world where moats and drawbridges are in short supply, how will you ever defend your turf? Or perhaps you want to expand your realm. *MiniWeapons of Mass Destruction: Build Siege Weapons of the Dark Ages* provides step-by-step instructions on how to turn everyday household and office items into 35 different medieval weapons for the modern era, including:

- ➲ Candy Box Catapult
- ➲ Chopstick Bow
- ➲ Bottle Cap Crossbow
- ➲ Clothespin Ballista
- ➲ Marshmallow Catapult

- ➲ CD Trebuchet
- ➲ Tic Tac Onager
- ➲ Mousetrap Catapult
- ➲ Plastic Ruler Crossbow
- ➲ And more!

Once you've assembled your arsenal, the author provides a number of targets to practice your shooting skills—an empty milk carton is converted into a siege tower, an oatmeal box into a castle turret, and more. Once you're armed and trained, there's no need for your desk, cubicle, or personal space to go undefended. Huzzah!

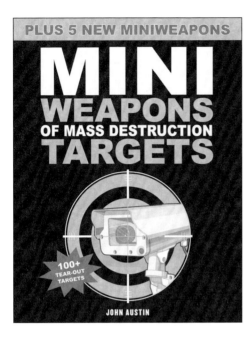

# MiniWeapons of Mass Destruction Targets

### 100+ Tear-Out Targets, Plus 5 New MiniWeapons

John Austin

978-1-61374-013-2
$9.95 (CAN $10.95)

The key to becoming an accomplished marksman is to practice, practice, practice. *MiniWeapons of Mass Destruction Targets* contains more than 100 tear-out targets to develop your skills. The targets are divided into three themes—Basic, Secret Agent, and Dark Ages—with a variety of gameplay scenarios. Blast the lock off a chained door, knock down a castle gate, compete in a game of Around the World, or shoot several miniature targets at various locations. Rules on the back of each target describe basic and advanced play.

In addition to the 100+ targets, MiniWeapons master John Austin provides instructions for building five new MiniWeapons perfect for target shooting:

➲ Paper Pick Blow Gun     ➲ Semiautomatic Pen Pistol
➲ Spitball Shooter with Clip   ➲ Pen Cap Dart
➲ Toothpick Tape Dart

Safety instructions are also included, as well as a guide to setting up an in-house firing range that will protect walls and furniture.

# So Now You're a Zombie

## A Handbook for the Newly Undead

### John Austin

978-1-56976-342-1

$14.95 (CAN $16.95)

Zombies know that being undead can be disorienting. Your arms and other appendages tend to rot and fall off. It's difficult to communicate with a vocabulary limited to moans and gurgles. And that smell! (Yes, it's *you*.) But most of all, you must constantly find and ingest human brains. *Braaaains!!!*

What's a reanimated corpse to do?

As the first handbook written specifically for the undead, *So Now You're a Zombie* explains how your new, putrid body works and what you need to survive in this zombiphobic world. Dozens of helpful diagrams outline attack strategies to secure your human prey, such as the Ghoul Reach, the Flanking Zeds, the Bite Hold, and the Aerial Fall. You'll learn how to successfully extract the living from boarded-up farmhouses and broken-down vehicles. Zombiologist John Austin even explores the upside of being a zombie. Gone are the burdens of employment, taxes, social networks, and basic hygiene, allowing you to focus on the simple necessities: the juicy gray matter found in the skulls of the living.

# Last-Minute Survival Secrets

### 128 Ingenious Tips to Endure the Coming Apocalypse and Other Minor Inconveniences

Joey Green

978-1-61374-985-2
$16.95 (CAN $19.95)

The Department of Homeland Security advises all citizens to develop an Emergency Preparedness Plan, along with a Disaster Supply Kit . . . but who has the time? What will you do if Hurricane Bernadette blows ashore before you can stock up on K rations and signal flares? Don't panic—it's Joey Green to the rescue!

*Last-Minute Survival Secrets* contains more than a hundred ingenious survival tips that may sound quirky at first but really do work. Green shows how to start a campfire with potato chips, open a locked suitcase with a ballpoint pen, and prevent heat stroke with a disposable diaper. Readers will learn to build a solar cooker using cardboard and aluminum foil, and a Wi-Fi antenna with a coffee can. It's the perfect resource for armchair survivalists, budding MacGyvers, and adventurists on a budget.

# Defending Your Castle

### Build Catapults, Crossbows, Moats, Bulletproof Shields, and More Defensive Devices to Fend Off the Invading Hordes

William Gurstelle

978-1-61374-682-0
$16.95 (CAN $19.95)

Your home is your castle, but could it withstand an attack by Attila and the Huns, Ragnar and the Vikings, Alexander and the Greeks, Genghis Khan and the Mongols, or Tamerlane and the Tatars? Engineer William Gurstelle poses this fascinating question to modern-day garage warriors and shows how to build an arsenal of ancient artillery and fortifications aimed at withstanding these invading hordes.

Each chapter introduces a new bad actor in the history of warfare, details his conquests, and features weapons and fortifications to defend against him and his minions. Clear step-by-step instructions, diagrams, and photographs show how to build a dozen projects, including Da Vinci's Catapult, Carpini's Crossbow, the Crusader-Proof Moat, and the Cheval-de-frise.

# The Practical Pyromaniac

**Build Fire Tornadoes, One-Candlepower Engines, Great Balls of Fire, and More Incendiary Devices**

William Gurstelle

978-1-56976-710-8
$16.95 (CAN $18.95)

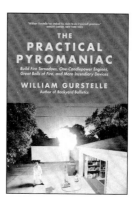

"What a fun, totally engrossing book! Gurstelle's projects—everything from a tiny single-candle engine to a flamethrower—are both easy to build and hard to resist. . . . Think of *The Practical Pyromaniac* as a cookbook for the budding scientist in each of us." —James Meigs, editor in chief of *Popular Mechanics*

*The Practical Pyromaniac* combines science, history, and do-it-yourself pyrotechnics to explain humankind's most useful and paradoxical tool: fire. William Gurstelle, frequent contributor to *Popular Mechanics* and *Make* magazine, presents dozens of projects with instructions, diagrams, photos, and links to video demonstrations that enable people of all ages (including young enthusiasts with proper supervision) to explore and safely play with fire.

# Backyard Ballistics, 2nd edition

**Build Potato Cannons, Paper Match Rockets, Cincinnati Fire Kites, Tennis Ball Mortars, and More Dynamite Devices**

William Gurstelle

978-1-61374-064-4
$16.95 (CAN $18.95)

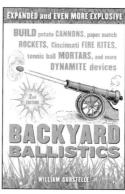

"How is it possible not to love a book with chapter titles like 'Back Porch Rocketry' and 'Greek Fire and the Catapult'? I devoured this prodigious account of all things explosive." —Homer Hickam, author of *Rocket Boys*

This bestselling guide has been expanded and updated, enabling ordinary folks to construct even more exciting ballistic devices in their garage or basement workshops than ever before. Clear instructions, diagrams, and photographs show how to build projects ranging from the simple—a match-powered rocket—to the more complex—a tabletop catapult—to the classic—the infamous potato cannon—to the offbeat—the Cincinnati fire kite. Four spectacular projects have been added to the fun arsenal: the spud-zooka, the powder keg, the electromagnetic pipe gun, and the sublimator.

## The Art of the Catapult

### Build Greek Ballistae, Roman Onagers, English Trebuchets, and More Ancient Artillery

William Gurstelle

978-1-55652-526-1
$16.95 (CAN $18.95)

"This book is a hoot . . . the modern version of *Fun for Boys* and *Harper's Electricity for Boys*."
—*Natural History*

Whether playing at defending their own castle or simply chucking pumpkins over a fence, wannabe marauders and tinkerers will become fast acquainted with Ludgar, the War Wolf, Ill Neighbor, Cabulus, and the Wild Donkey—ancient artillery devices known commonly as catapults. Instructions and diagrams illustrate how to build seven authentic, working model catapults, including an early Greek ballista, a Roman onager, and the apex of catapult technology, the English trebuchet.

## Soda-Pop Rockets

### 20 Sensational Rockets to Make from Plastic Bottles

Paul Jarvis

978-1-55652-960-3
$16.95 (CAN $18.95)

Anyone can recycle a plastic bottle by tossing it into a bin, but it takes a bit of skill to propel it into a bin from 500 feet away. This fun guide features 20 different easy-to-launch rockets that can be built from discarded plastic drink bottles. After learning how to construct and launch a basic model, you'll find new ways to modify and improve your designs. Clear, step-by-step instructions with full-color illustrations accompany each project, along with photographs of the author firing his creations into the sky.

# Gonzo Gizmos

## Projects & Devices to Channel Your Inner Geek

Simon Field

978-1-55652-520-9
$16.95 (CAN $18.95)

This book for workbench warriors and grown-up geeks features step-by-step instructions for building more than 30 fascinating devices. Detailed illustrations and diagrams explain how to construct a simple radio with a soldering iron, a few basic circuits, and three shiny pennies; how to create a rotary steam engine in just 15 minutes with a candle, a soda can, and a length of copper tubing; and how to use optics to roast a hot dog, using just a flexible plastic mirror, a wooden box, a little algebra, and a sunny day. Also included are experiments most science teachers probably never demonstrated, such as magnets that levitate in midair, metals that melt in hot water, a Van de Graaff generator made from a pair of empty soda cans, and lasers that transmit radio signals.

# Return of Gonzo Gizmos

## More Projects & Devices to Channel Your Inner Geek

Simon Field

978-1-55652-610-7
$16.95 (CAN $22.95)

This fresh collection of more than 20 science projects—from hydrogen fuel cells to computer-controlled radio transmitters—is perfect for the tireless tinkerer. Its innovative activities include taking detailed plant cell photographs through a microscope using a disposable camera; building a rocket engine out of aluminum foil, paper clips, and kitchen matches; and constructing a geodesic dome out of gumdrops and barbecue skewers. Most of the devices can be built using common household products or components available at hardware or electronic stores, and each experiment contains illustrated step-by-step instructions with photographs and diagrams that make construction easy.

## The Paper Boomerang Book

### Build Them, Throw Them, and Get Them to Return Every Time

Mark Latno

978-1-56976-282-0
$12.95 (CAN $13.95)

*The Paper Boomerang Book* is the first-of-its-kind guide to this fascinating toy. Boomerang expert Mark Latno will tell you how to build, perfect, and troubleshoot your own model. Once you've mastered the basic throw, return, and catch, it's on to more impressive tricks—the Over-the-Shoulder Throw, the Boomerang Juggle, the Under-the-Leg Catch, and the dreaded Double-Handed, Backward, Double-Boomerang Throw. And best of all, you don't have to wait for a clear, sunny day to test your flyers—they can be flown indoors in almost any sized room, rain or shine.

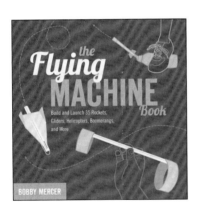

## The Flying Machine Book

### Build and Launch 35 Rockets, Gliders, Helicopters, Boomerangs, and More

Bobby Mercer

978-1-61374-086-6
$14.95 (CAN $16.95)

Calling all future Amelia Earharts and Chuck Yeagers—there's more than one way to get off the ground! *The Flying Machine Book* will show you how to construct 35 easy-to-build and fun-to-fly contraptions that can be used indoors or out. Better still, each of these rockets, gliders, boomerangs, launchers, and helicopters can be made for little or no cost using recycled materials. Rubber bands, paper clips, straws, plastic bottles, and index cards can all be transformed into amazing, gravity-defying flyers, from Bottle Rockets to Grape Bazookas, Plastic Zippers to Maple Key Helicopters. Each project contains a materials list and detailed step-by-step instructions with photos, as well as an explanation of the science behind the flyer. Use this information to modify and improve your designs, or explain to your teacher why throwing a paper airplane is a mini science lesson.

# The Hot Air Balloon Book

## Build and Launch Kongming Lanterns, Solar Tetroons, and More

Clive Catterall

978-1-61374-096-5
$14.95 (CAN $16.95)

More than a century before the Wright brothers' first flight, humans were taking to the skies in hot air balloons. Today, with basic craft skills, you can build and safely launch your own balloons using inexpensive, readily available materials. Author and inventor Clive Catterall provides illustrated, step-by-step instructions for eight different homemade models, from the Solar Tetroon to the Kongming Lantern, as well as the science and history behind them. *The Hot Air Balloon Book* also shows readers ways to heat the interior air that lifts these balloons, from tea candles to hair dryers, kitchen toasters to the sun's warming rays.

# Unscrewed

## Salvage and Reuse Motors, Gears, Switches, and More from Your Old Electronics

Ed Sobey

978-1-56976-604-0
$16.95 (CAN $18.95)

*Unscrewed* is the perfect resource for all UIYers—Undo It Yourselfers—looking to salvage hidden treasures or repurpose old junk. Author Ed Sobey will show you how to safely disassemble more than 50 devices, from laser printers to VCRs to radio-controlled cars. Each deconstruction project includes a "treasure cache" of the components to be found, a required tools list, and step-by-step instructions, with photos, on how to extract the working components. It also includes suggestions on how to repurpose your electronic finds. Fight the mindset of planned obsolescence—there's technological gold in that there junk!

## The Way Toys Work

### The Science Behind the Magic 8 Ball, Etch A Sketch, Boomerang, and More

Ed Sobey and Woody Sobey

978-1-55652-745-6
$14.95 (CAN $16.95)

"Perfect for collectors, for anyone daring enough to build homemade versions of these classic toys and even for casual browsers." —*Booklist*

Profiling 50 of the world's most popular playthings—including their history, trivia, and the technology involved—this guide uncovers the hidden science of toys. Discover how an Etch A Sketch writes on its gray screen, why a boomerang returns after it is thrown, and how an RC car responds to a remote control device. This entertaining and informative reference also features do-it-yourself experiments and tips on reverse engineering old toys to observe their interior mechanics, and even provides pointers on how to build your own toys using only recycled materials and a little ingenuity.

## The Way Kitchens Work

### The Science Behind the Microwave, Teflon Pan, Garbage Disposal, and More

Ed Sobey

978-1-56976-281-3
$14.95 (CAN $16.95)

If you've ever wondered how a microwave heats food, why aluminum foil is shiny on one side and dull on the other, or whether it is better to use cold or hot water in a garbage disposal, now you'll have your answers. *The Way Kitchens Work* explains the technology, history, and trivia behind 55 common appliances and utensils, with patent blueprints and photos of the "guts" of each device. You'll also learn interesting side stories, such as how the waffle iron played a role in the success of Nike, and why socialite Josephine Cochran *really* invented the dishwasher in 1885.

**Available at your favorite bookstore, by calling (800) 888-4741, or at www.chicagoreviewpress.com**

CHICAGO REVIEW PRESS